Securing the Outdoor Construction Site

Securing the Outdoor Construction Site

Strategy, Prevention, and Mitigation

Kevin Wright Carney

ELSEVIER

AMSTERDAM • BOSTON • HEIDELBERG • LONDON
NEW YORK • OXFORD • PARIS • SAN DIEGO
SAN FRANCISCO • SINGAPORE • SYDNEY • TOKYO

Butterworth-Heinemann is an imprint of Elsevier

Acquiring Editor: Tom Stover
Editorial Project Manager: Hilary Carr
Project Manager: Punithavathy Govindaradjane
Designer: Greg Harris

Butterworth-Heinemann is an imprint of Elsevier
The Boulevard, Langford Lane, Kidlington, Oxford OX5 1GB, UK
225 Wyman Street, Waltham, MA 02451, USA

ISBN: 978-0-12-802383-9

British Library Cataloguing-in-Publication Data
A catalogue record for this book is available from the British Library.

Library of Congress Cataloging-in-Publication Data
A catalog record for this book is available from the Library of Congress.

For Information on all Butterworth-Heinemann publications
visit our website at http://store.elsevier.com/

Working together
to grow libraries in
developing countries

www.elsevier.com • www.bookaid.org

Dedication

To William Wright Carney
1st Lieutenant, U.S. Army, WWII and Korea
And a hell of an engineer
Requiescat in Pace, Pater Meus

Contents

Introduction

I don't write like other people. I don't intend to impress anyone with lofty words or vaguely high-sounding concepts designed to make you think that what's left of my Irish brain is smarter than it really is. I'm old enough and ornery enough and have enough letters after my name to write the way that I think and speak.

What I offer is a down-to-earth assessment of what I see as an extremely costly nuisance that plagues every country in the world where there is construction of large outdoor capital projects, that is, roads, freeways, bridges, overpasses, runways, and just about anything that is made of concrete, steel, and other expensive stuff and that is designed to sit outside and last a very long time. This also includes concrete batch plants, government-sponsored concept projects, gravel yards, and gravel mines. The costly nuisance that I speak of would be theft of everything from steel and copper, to piles of gravel and sand, to huge pieces of heavy machinery equipment. Although some statistics exist, the cost of these thefts is nearly impossible to calculate worldwide, because it goes beyond the cost of the materiel and equipment that is stolen. It extends to the lost profit and construction delays when vital materials and machinery are missing at the beginning of a construction day.

An assessment is virtually worthless without suggested remedies, which I will proffer in this book. The remedies that I suggest are what I see as logical solutions to this theft epidemic and for the most part can be applied worldwide. They are not the only possible answers. Everyone solves problems based on their unique background and experience, and if you come up with a better solution, then by all means, go for it. Just think it through first, hopefully using some of the principles in this book, so that it works for your site.

My background began as a law enforcement officer and leader, with more than 23 years of experience before I retired. You would think that with that kind of experience, I would have arrested hundreds of construction thieves over the course my career. The fact is that I never arrested anyone, even once, for stealing from a construction site. Does that mean that I was a lazy cop? Not hardly. The fact as I see it looking back was that I didn't know what to look for and may have driven by crimes in progress that I didn't recognize as criminal activity. But I will get into that as the book progresses.

I did not even think about theft from construction projects until I started my second career in the security business. Even then I didn't think about it much until I became the general manager of a security and investigations corporation, which, among other things, handles the security for large outdoor construction projects. I learned that anything can and does happen in the construction industry, and if you are not flexible and prepared, you'd better be prepared to be responsible for some hefty but preventable losses.

I have outlined information and strategies here that should be helpful to every reader who buys this book. I hope you find value in it. If you have any involvement in construction, then this should be a good read.

Kevin Wright Carney, CPP BA, OSJ
Sergeant (Retired), Los Angeles Sheriff's Department
Member ASIS International and Certified Protection Professional
Newbury Park, California, USA

> As **a note**, if you are in any way involved in securing outdoor construction projects, I highly recommend joining **ASIS International**. This worldwide organization is composed of tens of thousands of professionals from all fields who come together to share ideas and to provide education and training in the security field. Not only will you increase your knowledge of the security field, but you will also meet some really nice people and attend some great events.

Who Should Read This Book

1. **Security professionals:** Our clients are paying us with money that is taken away from their bottom line. We owe it to them to prove that every dime they have paid us to provide security services represents a positive return on their investment. You will find anecdotes from which you can begin to think of ways that you and your security team could have prevented a loss or could prevent a future loss. You will also find strategies and solutions to these problems as I see them. These strategies are not exhaustive but should get you to look at these sites as more than just placing a live scarecrow with a uniform on site or an automatic camera on site, and thinking that is sufficient.

   Wherever possible, I will provide you with statistics. I was surprised to learn that there seems to be no central repository for construction site thefts statistics. However, statistics mean very little to the clients when they experience a loss. They will look to you for answers, not excuses, and you had better have them if you want to keep the contract.

2. **Construction professionals:** You want to get outdoors and build your stuff. I get that. You don't have time to worry about security. I want to convince you to stop rolling the dice and to think about security at the beginning of every project. The time you spend planning and implementing site security will save you hours of aggravation later and will save you money. Preventing thefts will help you to meet or beat deadlines by avoiding costly delays.

   In this book, I want to convey to you the importance of planning on attacks by thieves and give you mitigation and prevention strategies. I will also discuss the choice of providing your own security versus hiring a security services contractor.

3. **Law enforcement professionals:** If you joined law enforcement as I did, to catch crooks and take them to jail to protect the people we serve, then you should get familiar with the contents of this book. I guarantee that at some time in your career you have driven by or will drive by a major crime happening in plain sight right before your eyes, without recognizing what you have just seen. Whether you are an executive, a manager, a supervisor, a detective, or a patrol officer, you need to make yourself aware of the contents of this tome. You may find yourself at the head of investigating and handling a great caper one day that will go beyond the day-to-day criminal investigation.

   You will learn that theft from construction sites goes beyond the theft of metals, materiel, heavy equipment, and trucks (lorries). Although costly construction delays cannot likely be criminally charged to a perpetrator, they can and do represent a significant dollar loss to the construction professionals whom you serve. In these days of proactivity, crime prevention through awareness and diligence may very well keep crime statistics in your jurisdiction at a lower rate and thus contribute to the overall effectiveness of your agency.

4. **Insurance professionals:** Your profession most likely carries the heaviest financial burden of these construction thefts, reducing your company's bottom line through the payment of huge claim settlements. You, perhaps more than anyone else, can impact attitudes about construction security. Your profession is responsible for setting insurance rates, thus impacting the bottom line of every construction project. Through improving rates for construction companies that clearly demonstrate a proactive security plan, your industry can impact construction site theft mitigation and save you money.

5. **Legal professionals:** Your profession has perhaps the broadest range of concern about the problem of construction theft and other serious incidents at construction sites. As lawyers, you are the prosecutors, the defenders, the litigators, the judges, and the legislators whose actions wield the hammer of justice against these crimes and their perpetrators. You are the physical embodiment of the law, and none of the actions of law enforcement or insurance professionals can be enforced without your expertise and efforts. I hope this book will trigger ideas for you as you prepare your cases or write laws to help mitigate the effects of this serious problem. Solutions such as sentencing enhancements for thefts from construction sites will help immeasurably to stem the impact of these crimes.

About the Author

Kevin Wright Carney, CPP, BA, OSJ, is the President and CEO of Silver Gauntlet International LLC, a security consulting firm that provides vulnerability assessments, site surveys and security plans primarily for construction companies (www. silvergauntlet.com). He is also a retired sergeant with the Los Angeles Sheriff's Department, having served more than 23 years.

He is an active member of the American Society of Industrial Security (ASIS International) and a board Certified Protection Professional (CPP). He assists as a mentor for the annual Southern California CPP Course. He is also a member of the Association of Threat Assessment Professionals (ATAP), and belongs to the Los Angeles chapter.

He has a bachelor's degree in political science with an area of specialization in foreign government from California State University, Northridge. He speaks five languages and is an accomplished choral musician and artist. He is also a Knight of St. John.

He is a major in the Civil Air Patrol and has been a member for nearly fifty years, having joined as a cadet in 1966. He is a certified SCUBA diver, and restores antique license plates in his spare time. He also collects and studies water turtles. He writes both fiction and non-fiction books.

He has three grown children and three grandchildren. He lives in Newbury Park, California, near the rolling hills, just inland from the great Pacific Ocean.

ALSO BY THIS AUTHOR
A COLD NIGHT IN THE ATLANTIC

Deep in the dark Atlantic, the U.S. Navy's newest submersible *Ballard* is on a top secret training mission at the wreck of the *RMS Titanic*. In the stern half of the wreck, Commander Joseph R. Browder discovers a huge cache of gold coins and a secret that will rock the world.

Travel back to 1912 with the crew and the shipbuilders and relive the dark conspiracy of gold, dynamite, and murder. This is an exciting adventure that you will want to read cover to cover, nonstop. There is so much more about this tragedy than has ever been told before.

Kevin Wright Carney puts you aboard the ill-fated liner in a way that you have never experienced. You will never look at the wreck of the *Titanic* the same way again.

THE BEAST OF THE ANGELES: THE ANGELES CREST MURDERS

A brutal serial killer is loose in the Angeles National Forest. He is savage, ruthless, and unstoppable. No camper, skier, or hiker is safe.

Los Angeles County Deputy Sheriff Kent Wickham, who patrols the forest alone, is responsible for finding and stopping this vicious killer. He not only must battle a murderer but also must fight elements within his own department to restore order to the most magnificent place in Los Angeles County.

Kevin Wright Carney takes you inside a Los Angeles Sheriff's Department radio car and inside the life of a deputy sheriff. You will understand police work in a way that you never have before. You will be with the Sheriff's Department and join their quest to stop a human beast, the Beast of the Angeles.

Acknowledgments

Anyone who has written a book, even a novel, knows that there are people whose help was indispensable. Brian Romer of Butterworth-Heinemann publishers was one of those people. To this day, I don't know how he found my name or found out that I write books. I pitched the concept of this book to him and explained my reasons why this subject is important worldwide, and he, after giving the idea some consideration, agreed to let me write this book. Also, Keira Bunn of the same company has been of immeasurable help in making sure that I got everything done correctly. Hilary Carr took over as I was completing the book and helped me with deadline extensions and paving the way for production. She devoted a lot of time and positive energy to this effort. Thanks to them all.

No writer can write well without having someone look at and evaluate his or her work as it progresses. I must thank my progeny team, Travis Carney, Carleen Littig, and Clint Carney, for reviewing this book for grammar, punctuation, and syntax. Thank you all, Travis, Carleen, and Clint. Travis and Clint gave me the grammar drubbing that I needed to help all of this make sense. Travis helped by making sure the checklists and worksheets look like they were designed by an adult.

My friend, Sheriff's Academy classmate, and retired L.A. Sheriff's Lieutenant Paul Scauzillo did a chapter-by-chapter review and helped me when I got bogged down in my own verbiage. Thanks, Paul.

As I have mentioned, my experience lies in law enforcement, and I needed the expertise and knowledge of some of those people who devote their lives building these magnificent structures that we all seem to take for granted. Jason Mattivi of Security Paving in California first comes to mind. He and his brothers own and run the company that was started by their father. By watching and listening to him, I developed and appreciation for the passion that construction professionals have for their projects, no matter how large or small. Mark Christie, Gary Baxter, and Hani Jamaleddine of Security Paving also answered a lot of questions as I was writing this book.

Next in that group of people is Dave Huestis of Siemens International, whom I met while bidding on an AQMD concept project. He helped me at the beginning of the book with some concepts that would make the book appealing on an international level.

One of the nice things about belonging to ASIS International is that you get to meet a whole lot of people with outstanding backgrounds who are willing to share their expertise if you ask them. One such person is Pamela Graham, retired Supervisory Special Agent with the FBI. I asked her to review my sections on employee screening and background investigations. She gave me a detailed analysis of what I had already written and pointed me to a host of concepts that I had not considered. Basically, she helped me make this book better. Thanks, Pam.

Jim Grayson, CPP, a friend and fellow author, is a retired law enforcement manager (sworn) and an expert on vulnerability and site assessments. He always took my phone calls and offered cogent commentary on the many directions that my mind followed while writing this book. Thanks, Jim.

I reached out to the National Equipment Register (NER) for further information on their statistics. Elizabeth Ohanyan, Marketing Analyst, responded with information and permissions, and suggested that I contact Giuseppe Barone, a headquarters manager, and schedule an appointment to speak with Ryan Shepherd, the General Manager. Ryan Shepherd gave me as much time as I needed for the interview, and provided me with an abundance of information on the NER and how it works to benefit the construction and insurance industries, as well as law enforcement. Thanks to Ryan, Giuseppe and Elizabeth, this book is more complete.

Anyone else whom I forgot, please forgive me for not acknowledging your help.

Kevin Wright Carney, CPP, BA, OSJ

The Impact of Construction Site Theft

1

This chapter outlines the costs in terms of time lost, equipment and materiel stolen, danger to site personnel, insurance premiums, and loss of reputation and introduces readers to some basic concepts regarding the securing of outdoor large capital construction projects.

Theft from construction sites is probably older than the pyramids in Egypt. As long as there has been something to steal, thieves have been stealing the stuff that is needed to build the structures that make the civilized world a great place in which to live. An entire book could be written about the thieves, and to be sure, we will talk about them, but this book is about how all of us in the security, construction, law enforcement, insurance, and legal professions can put our collective knowledge together to mitigate construction site thefts, if not stop them altogether.

Any construction professional can give you a very close estimate of how much any one security incident costs his or her company. Those of us who are not in the construction business tend to think of these thefts in terms of the cost of the machinery and materiel stolen alone. Most certainly, these costs are astronomical and represent the financial impact around which we can begin to wrap our brains.

However, there are costs beyond replacing that which is stolen, which branch out from the initial incident. These losses include loss of construction time, which has a variety of impacts. For instance, if a construction project has a rush deadline and major vital pieces of equipment or materiel are stolen and the construction managers cannot work around that theft, the delay could cost the construction company significant money in early-completion bonus monies.

Additionally, as in all industries, reputation is the key to future business. Repeated preventable completion delays over time can damage the reputation of an otherwise great company and eventually cost future contracts. These lost contracts translate into millions of dollars left on the table for someone else to pick up.

There is a human component to this dilemma as well. After a major theft, companies must either contact workers in advance of their shifts and advise them not to show up for work or send them home after they have arrived on site. This costs the company money in wages paid for work not performed and impacts the livelihoods of the people who depend on this work to provide for themselves and their families.

One other human component occurs when the personal tools of construction workers are stolen. This represents a major loss to the family income of such workers and is often not covered by insurance.

In some cases, construction companies can either rent equipment, if available, until the replacement equipment can be found, or use other pieces of equipment to perform a task for which they were not intended. Either way, this is not optimal, and although it may somewhat mitigate the situation, it is costly nevertheless.

Insurance rates may not rise after a preventable construction theft loss because policies generally give blanket coverage for the insured company, but certainly the insurance industry looks at the overall risk for a given area. Similar to all businesses, insurance companies are in business to make money rather than lose it or break even. Discounts are offered to construction clients who take reasonable steps to prevent or mitigate theft. These discounts increase a project's bottom line, and frankly, a good security plan that is implemented is an inexpensive additional layer of insurance. One of the purposes of this book is to change the mindset of construction company executives to look at a security program as an essential part of the overall job plan. Why wait until you suffer a significant loss to secure your site?

According to statistics from the LoJack Corporation and the National Equipment Register (NER), theft of construction mechanized equipment costs between $400 million and $1 billion every year in the United States alone. That cost is for the machinery only and does not cover the cost of materiel, cables, and tools plus the cost of lost construction time. However, it is interesting to note from the LoJack and NER statistics that 83% of all construction companies have experienced theft, and 64% of those companies never got their equipment back. Also, 46% of the stolen equipment is 5 years old or newer, which tells us that the thieves are filling orders by request.[1]

This book is intended to be international in scope, but it is interesting to note, as mentioned earlier, that there is no central repository for global statistics for construction site theft. Just like the news, all crime is local, and thus, the statistics are kept by the jurisdictions and private organizations closest to each reader. Most statistics are broken down by what is stolen and not specifically by locations, such as construction sites.

---

[1] Statistics courtesy of the LoJack Company.

I will use anecdotes throughout the book to illustrate points. Many of the anecdotes in this book will be from Southern California, simply because that is where I live and work. If you would like to make the anecdotes more personal, then by all means, erase the California name and fill in the name of a site near to where you live. You won't hurt my feelings. The principles are the same, wherever in the world that you happen to live.

To better understand my anecdotes, you should know that I was the general manager of a medium-sized security and investigations corporation that operated throughout California, in the United States. Among the security services that we provided, we furnished security officers for large outdoor capital projects throughout the state. It was in this capacity, coupled with my law enforcement background, that I came in contact with the construction industry and developed a passion for devising strategies for keeping these sites free from thieves and vandals.

THE INTERSTATE 5 PROJECT IN NORWALK: PART ONE

One morning early in 2014, I got a call from Jason Mattivi, one of the owners of Security Paving, a major California construction company that builds all manner of freeway, bridge, and other concrete projects. He asked me to come and look at the beginning of a freeway expansion project off the Interstate 5 Freeway in Norwalk, which extended several miles.

When I arrived, I met Jason on a side road beside the freeway, which was filled with traffic rushing by. We were below the freeway where a series of cul-de-sacs in a residential area had been graded away to make way for the building of a huge retaining wall, behind which would be filled with material, and upon which the new freeway lanes would be built.

We exited our respective vehicles, entered an opening in a chain link fence, and began walking the length of one side of the project. He began telling me of the rebar and framing lumber that had been stolen from the site in the past few days and asked me if I could secure it with patrols of the concerned area. As we walked from street to street through the graded dirt, I saw a stack of iron rebar lying on the ground; the stack was about 3 ft high by 3 ft wide (about 1 m high by 1 m wide) (Figure 1.1).

Keeping in mind that construction sites in Southern California are somewhat similar to the ever-present mushrooms in a dense forest, I casually asked him how much the stack of rebar cost. To my shock, he told me that that each such stack cost $60,000, and there were dozens of these stacks lying about. That was my "aha moment" for the outdoor construction business. It was in that moment I realized that construction companies throughout the world actually leave piles of cash (in the form of construction materials) lying about their construction sites overnight and through the weekends, trusting to fate or the stars that it will be there in the morning when they are ready to begin work.

FIGURE 1.1

$60,000 lying on the ground.

THE BRIDGE AT HESPERIA: PART ONE

On a March morning in 2014, I got a call from Mark Christie at Security Paving in Sunland, California. A theft had occurred from the site of a freeway overpass bridge under construction on the Interstate 15 Freeway near Hesperia, California (Figure 1.2), and he needed us to set up a security patrol at that project. This site was a remote area that was unlit at night.

As is proper, I responded to the site to survey it in person. (No security professional should ever set up a post like this without first looking at it with his or her own eyes.) Upon arrival, I met with the site manager, Gary Baxter, who advised me that they had experienced theft of batteries, cables, and equipment from construction vehicles and that they wanted to set up a patrol service to prevent this type of theft from recurring.

This bridge construction site was in the beginning stages. Basically, the project was to build an overpass across the Interstate 15 freeway. This bridge was to span a 10-lane freeway with a huge median that could have easily accommodated four additional lanes. The earthen inclines that would support the bridge and its on and off ramps had only begun to be graded. From the east side of the freeway, I could not see where the west side was starting because of the distance between both sides of the bridge. This was to become a major problem for the first few months of the project. Added to that, it was a 2-mile drive to get from one side of the project to the other via side roads and other freeway bridges.

FIGURE 1.2

The bridge at Hesperia from the upper west.

Company planners wanted to secure the site by use of one security officer stationed on one side of the freeway, convinced that the officer would be able to see anything on the other side of the freeway, even when it was dark out. This was to prove to be a costly decision. I explained to Mark and Gary that at the very least, we needed to patrol both sides, but this was decided to be too expensive. I say this not to unfairly cast blame on the client because they are both great construction professionals but to illustrate that the security budget eats into the project's bottom line unless security costs are built into the project's budget by the buyer at the time of the Request for Proposal (RFP).

We will come back to this bridge later.

DANGER TO SECURITY PERSONNEL AT THE SITE

A statistic that is most likely never tracked is the occurrence of after-hours assaults to the people left behind to secure the construction site. The men and women whom we employ at these sites have an unglamorous job. In years past, there was an image of a night watchman, who was generally an older person who was finishing out his golden years by keeping an eye on things and who would notify the police when something suspicious arose.

Today's men and women who perform these important duties are somehow deemed to be at the bottom of society's workforce, and the term "security guard" has almost become a pejorative. Don't ask me why because I see these people every day, and for the most part, they are hardworking, observant, and diligent. They work hours that most people would avoid and receive little praise outside of the companies for whom they work.

At our construction sites, we leave these people in remote areas all alone with only a flashlight and a radio or a telephone with which to call for help if they need it. Both security professionals and the construction professionals who hire them should pay close attention to the size, bearing, and abilities of the people whom we are placing in danger at these sites. Make sure that your company's central dispatch office keeps track of them at regular intervals during the night. Make sure that both dispatch and the officer have the telephone number to the nearest law enforcement agency to the project and make sure that the account manager has made a liaison with local law enforcement at the outset and during the life of the project.

During my years as a law enforcement officer, I learned that every year, around four to six police officers are killed in the United States alone because of falling asleep on duty in the nighttime and being shot to death by criminals. Sadly, no such statistics are kept on security officers, but the likelihood of nighttime assaults on lone security officers in the middle of nowhere compels us to develop strategies to keep officers awake during the night. We will further discuss these strategies later. It must be kept in mind that whereas equipment and materiel that is stolen can be replaced, the lives of our security officers cannot be replaced.

Always keep in mind that the people who steal from construction sites are criminals. They know that in most countries, thefts of this magnitude carry heavy penalties. To them, the life of one security officer means little compared with the years that they may spend in prison if they are caught. Apparently, the penalties are not doing enough to deter criminals from these profitable crimes.

YOU GET WHAT YOU *IN*SPECT, NOT WHAT YOU *EX*PECT

The following example can be applied to any human endeavor under the sun, but I am directing this to security executives and managers in hopes of bettering our service to our clients.

The account managers and field supervisors who work for me hate this saying when I bring it up at nearly every staff meeting, but this old saw is true. **Carney's rule number 1: You get what you *in*spect, not what you *ex*pect.** An executive or manager generally forms a clear vision in his or her mind of how a project is to be secured and most times goes through the process of committing the idea to paper as a plan.

The vision is clear as glass, and after having given their input, everyone around the planning table nods their heads, acknowledging that their plan is the way to proceed. Several weeks later, you get a phone call from the construction site manager, asking you just what in the world is going on with your security "professionals" and telling you in the next breath that there has been a theft or vandalism or some other preventable catastrophe.

So, dutifully, you resurrect your carcass from your imperial executive chair, get in your vehicle, and respond to the site to inspect the plan that you had so brilliantly designed. Upon arrival, you discover that the only resemblance between the plan that you had devised and reality is that people in your company's uniform using your company's equipment are at the site, scratching their collective posteriors and wondering what just happened. Why? ***Because you did not go out and inspect when the project was first set up, nor did you cause routine inspections to be made during the course of the project.***

Some of us are lucky enough to have subordinates who can be trusted to accurately set up and execute a plan. However, because we do not and should not hire robots, people tend to alter plans to fit their own experiences and ideas of how things should be done. This is not necessarily a bad thing if it works. However, when things go belly up, you have no one to blame but yourself if you have not conducted regular site inspections or at least caused those inspections to be made with the results reported to you.

By periodic visits to a security site, you can ensure both that your vision is indeed a workable plan and that the people who work for you are paying attention to what is going on. This does not mean micromanaging, which grows exponentially ineffective as your company increases in size. However, through proper training and discussion, you should have effective communication with your account managers and field supervisors. When this relationship is developed and your subordinate managers know that their opinions are fairly weighed, you are much more likely to have the right information to evaluate your security plan and *make changes if necessary.*

The other added benefit to site visits is to develop a strong partnership with your clients. Keeping in mind that your clients are always busy, you should make the visit short but long enough to listen to the client and receive feedback to keep your security plan viable. Be ready to walk with the person as he or she goes about his or her duties. Developing that relationship with your client will help you when something in your brilliant plan goes wrong, which most likely will happen.

If you don't take the time to get in the field, you will be permanently enrolled in the M.B.A. program, that is, ***management by apology***.

CHAPTER POINTS BY DISCIPLINE

1. **Security professionals:** Begin to think of construction site security through the eyes of the client. Remember that every time there is a theft incident, it costs your client more than just the cost of the items stolen; it also leads to costly delays and possibly the loss of significant bonus monies. Keep in mind the human costs to your client's personnel when a theft loss occurs and the potential human cost to your security officers if your company loses the job because of theft losses.

    Remember that the quality of your site security plan and its execution represent an additional layer of inexpensive insurance to your client. Learn the value of the equipment, vehicles, and materials that your client has on site. When you add up the costs, you will begin to see how important it is to have the right plan and the right people.

Always be mindful of the safety and well-being of your security officers, to whom you entrust these construction sites. Remember to always provide them with adequate communication devices, weather-related uniform items, and safety equipment. Always make sure that they are able to escape danger if a site is attacked by thieves. Have a plan for regular safety checks during the night to keep them alert and to ensure their personal safety. Know the telephone number to the law enforcement agency at their location.

Remember that you get what you *in*spect, not what you *ex*pect. Don't be the one who subscribes to the MBA program.

If you are not already a member, join ASIS International as soon as you possibly can. You will meet other professionals from the security industry as well as other disciplines and will learn so much more about our industry than you ever thought possible. There are countless seminars and conferences where you can gain valuable information and network with so many professionals from a variety of disciplines. I guarantee that you will be happy that you joined.

2. **Construction professionals:** Always remember that proper security of your construction site can save you endless aggravation and untold thousands of dollars in delays, replacement of vehicles and equipment, and loss of reputation. Try to get your clients to include funds for security in the RFP rather than letting the cost come out of your bottom line later.

    Keep in mind that the amount of money that is lying about on the ground at your site in the form of materiel and equipment. Money is the lifeblood of business. Ask yourself if you would leave that much cash lying on the ground and expect that no thief would take it.

    Take some of that all-important time out of your workday at the beginning of each project to meet and confer with your security provider about the site security plan. It will reap untold benefits in the long run.

    If your company is large enough, designate one of your managers to join ASIS International to maintain a liaison with professionals in the security industry. Your representative will most likely meet and network with other construction professionals and learn new ways to mitigate or eliminate construction site theft.

3. **Law enforcement professionals:** No matter what level of the food chain you are on in law enforcement, if you are involved in any way with investigating crimes at construction sites or even if you are a patrol officer whose duties take you by construction sites, take the time to meet with the construction site manager and learn what is going on there and what you can do to prevent crime. If you are contacted by a security manager about a project, take the time to meet with that person and discuss how your agency can work hand in hand with the private guard service to prevent crime or to make a decent arrest. Try to keep your attitudes about private security open. You may learn that the manager is

a retired police officer or military person or has been in the security industry for years. You will learn that the security industry has progressed well beyond living scarecrows with plastic badges.

Familiarize yourself with organizations such as ASIS International, and if duties permit, make the time to join or at least attend some meetings. You may be pleasantly surprised at the many developments in the security industry.

4. **Insurance professionals:** If you are in any way involved in insuring outdoor construction sites, whether it is in setting rates, interacting with clients, or investigating loss incidents, familiarize yourself with the contents of this volume and the inner workings of the security industry. Keep in mind that the best claim incident possible is the one that never happens. Encourage, cajole, or insist that your construction industry clients have a security plan, designed by a security professional. Research the security companies in your area and vet them as a service to your clients. Find out which companies are professional and which provide the most professional, reliable service.

Find and join your local chapter of ASIS International. You will find that many of its members are not solely from the security industry. You will find members from every discipline imaginable, and you will learn many of the avenues to gain a clear insight into how to protect your clients' construction sites and improve your bottom line.

5. **Legal professionals:** Your profession has perhaps the broadest outside impact on construction site security. Through legislation, litigation, and prosecution, the legal profession guides how business is done across all lines of every industry. The potential of legal professionals to insure that proper security is maintained at every outdoor construction site is great. Consideration by prosecutors, judges, and legislators must be given to encourage sentence enhancements for thefts from construct sites to make the penalties for this type of crime so great that fewer criminals will want to attempt these thefts. In many jurisdictions, the punishment is generally no greater than any other grand theft case, yet the impact of construction site theft represents billions of dollars worldwide every year.

As with the other disciplines, I encourage you to join ASIS International. You will find that a large number of your colleagues already belong. Whether you are a civil or criminal attorney, prosecutor, judge, or legislator, you will gain a greater insight into all types of security requirements and procedures. You will find a rich source of materials and information to prepare or judge your case, or to write stronger legislation to help protect the construction industry from this insidious and costly problem of construction site theft.

What Is Being Stolen?

2

This chapter showcases the nature of theft at construction sites, detailing the types of thefts that are occurring, from the very small to the very large. It covers theft of machinery, tools, steel products, copper, steel, gravel, sand, and a wide variety of other things that are being stolen, as well as their worldwide statistical significance.

MACHINERY AND VEHICLES

Most people outside of the construction industry would be surprised to learn that there are no government titles for most construction equipment, except cars trucks and trailers. Now I'm not an advocate for more government regulation or oversight, but it seems unfathomable that there would be no official government agency that registers these machines. Later in this book, we will discuss national equipment registers, a concept that began in the United Kingdom. This is a step in the right direction, but so far it does not carry the weight of law.

Although the types of equipment stolen vary slightly from country to country throughout the world, they pretty much mirror each other. For instance, in the United States, light trucks are the most frequently stolen pieces of equipment, but in the United Kingdom, trailers are most likely to be stolen. Regardless of frequency of theft, any types of vehicle and equipment stolen from a project are always problematic.

A recent trend in the United Kingdom has occurred when thieves invent an imaginary construction site, rent construction vehicles and equipment, and have them delivered to what appears to be a legitimate new construction site. After all of the rented equipment is delivered, it is then transported away and sold on the black market. Although this kind of theft appears to only affect the rental business, it can also impact a construction company when needed machines are not available for rental because they have been stolen.

Of course, the fault of these thefts always lies with the thieves themselves and the unscrupulous people who buy the stolen vehicles and equipment, but steps can be taken to mitigate these thefts or prevent them by "hardening the targets."

What may surprise someone outside of the construction business is that frequently, construction companies have either poor or no records of their equipment and vehicles. It is not uncommon for a site manager to tell the security professional, insurance investigator, or law enforcement officer who is investigating a theft that he or she does not even know the make and model of the equipment and has not a clue as to what the serial number of the item is. When asked if someone at "the office" might have the information, the answer is often, "I don't think so."

Part of this is because of the mindset of "I just want to complete my project. Don't bother me with the details." Another part of this equation lies in the fact that many pieces of construction equipment do not use keys, or they have "universal" keys that fit every piece of equipment that a manufacturer makes. To further exacerbate the problem, many pieces of construction equipment, even motorized ones, do not have a serial number or Vehicle Identification Number (VIN). Believe me, tending to the clerical details of your equipment will pay off in spades. If you do not have the time to do it yourself, then hire a bright part-time clerk to attend to this important matter. It will save you a lot of money in the long term. The clerk's salary will be paid by the cost of equipment that is recovered.

MITIGATING STRATEGIES TO PREVENT VEHICLE AND EQUIPMENT THEFT

Face it. Many thefts occur because the criminal is organized and we are not. Construction thieves often come from the construction industry and already know the ins and outs of how things are done. Their organization may be poor and haphazard, but when compared with someone who is completely unorganized, they win the easy victory. **Remember, even a simple plan is better than no plan**.

1. **Keep records.** At a place other than the construction site, records of every piece of equipment that the company owns or has rented should be kept in an **organized** filing cabinet. This means that some employee, preferably not a construction worker, should set up files for every piece of equipment. These files should contain the receipts for the piece of equipment, the description and manufacturer, the model number, the year of manufacture, and the serial number if it has one. Included in these files should be a reference as to who the insurance carrier is and how to contact that carrier in the event of loss.

   A separate set of files should be kept for all pieces of rented equipment. These files should contain a copy of the rental agreement plus copies of any extra insurance purchased for that equipment. (If you look at your rental agreement, you will see the type of information that you should be capturing for the equipment that you own. You can copy their format to organize your own data.)

This information, when provided to law enforcement, will expedite the return of your equipment to you. Without this information, it may be impossible to prove that that stolen property belongs to you.

2. **Mark your equipment.** As previously mentioned, many pieces of equipment have no serial numbers or have serial numbers that are easily removed, (i.e., a number that is bolted to the equipment). You should develop a simple numbering system for your equipment that is unique to each piece. For example, if you own the XYZ Construction Company, you could use such a simple number system as XYZ 001, XYZ 002, XYZ 003, and so on. Then you should place these serial numbers in several places on each piece of equipment. Many criminals are lazy and will stop looking for serial numbers after they have removed or destroyed the most obvious ones. Place the other serial numbers in places that are not easy to discover, such as under inspection plates or along the bottom inside of the equipment's frame. If you have stamping tools or engraving equipment, I suggest you use those rather than labels that can be easily removed. In the United Kingdom and United States, you can get information and help in marking your equipment from the National Equipment Register.

Also, stencil your company's name in epoxy paint all over the equipment. Make the crooks work hard to remove your identity from your things. By having your name prominently marked on your machinery, it will make it that much easier for law enforcement to spot your equipment if it is stolen. If the theft is reported promptly, having your company's name displayed may ensure a quick return of the stolen items.

Even rental vehicles and equipment can bear your company's name by use of magnetic temporary signs, which can be easily purchase at any sign shop.

3. **Photograph your equipment.** You will probably see this a few more times in this book. Photographs are invaluable in describing what has been stolen to a law enforcement officer, a security professional, an insurance investigator, or an attorney in a case. Many of those people might not know what a skid steer loader, a walk-behind power trowel, or a feller buncher is, yet they will be responsible for identifying the stolen item and perhaps recovering it and identifying or prosecuting the offender(s) who stole it.

Looking at this scenario as a former law enforcement officer, I can say that knowing what a stolen object looks like greatly enhances the chances of recognizing it during the stop of a theft suspect.

In today's world of digital photography, a construction site manager can recall a photo from the company's database or his or her cell phone and send it to the responding officer's cell phone. That officer can, in turn, forward it to the cell phone of every law enforcement officer in the area, thus instantly widening the search area just by knowing what the equipment looks like.

4. **Secure your equipment.** This concept seems to be so basic as to not merit mention, but I can assure you that there are plenty of construction professionals who just turn off their equipment at the end of a workday and leave it, like a dropped teddy bear. Needless to say, a thief has only to wait until the jobsite is deserted, and he will be able to quickly abscond with whatever equipment or vehicles he wants. Why? Because **no one secured the equipment**.

It seems odd that early in the morning of every work day, a fueler arrives to make sure that all vehicles and equipment are fuelled and ready to go, yet in many cases, no one is assigned to secure to jobsite's vehicles and equipment at the end of the day.

Secure fencing is normally seen at every construction site. In that enclosure are usually found the office trailer, a storage shed, several light pieces of equipment, and perhaps some as-yet undeployed building materials. All too often at the end of the day, the gate, if there is one, is left unlocked or "secured" with a piece of chain or wire, which creates the illusion of security to honest people but does little or nothing to deter thieves. If that's all you are going to do, why bother?

Good fencing and properly locked and secured gates may only serve to delay a thief from his objective, but every moment that he is on the site increases the chance of discovery, and that is a good place to start. Now, keeping in mind that time is one of a thief's enemies, let's look at some other ways to deter theft.

Good lighting at night is another enemy to a thief. Light towers and the generators that they require are expensive, and the cost to keep them running eats into the bottom line of your project. However, darkness at night is a powerful ally for a criminal. If your yard or outlying equipment is unlit at night, a police officer could drive by every 10 minutes and not see criminal activity because the thief is working in the dark.

Cameras that can be monitored remotely and that record events are a great tool both forensically and in the moment if they are monitored in real time. Cameras are discussed in further detail in the Chapter 8.

Effective parking is a good way to deter theft when combined with good lighting. By parking your vehicles near and facing the closest travelled roadway, side by side, you increase the chance of thieves being seen by people who are using the road during the night, especially the police.

When parking equipment-hauling trailers, remove the equipment from the trailer and park it on the ground away from the trailer. By doing so, you increase the time that is required for a thief to complete the theft, and as said before, time is an enemy to a thief.

Wheel locks are an effective deterrent to vehicle and trailer theft (Figure 2.1). Even though eventually they can be defeated, they burn precious minutes off of the thief's time clock. Using wheel locks on everything with round wheels may cause thieves to go to another construction site where things are easier to steal.

Security officers are the "human element" that provide the convergence between all of your gadgets and the people in charge of your project. They are

FIGURE 2.1

A good wheel lock will delay the thief's time on site.

very important because their highly visible presence can deter a thief or alert law enforcement to a theft in progress. Many construction professionals choose security officers as a last result after a theft has already occurred. Often they look at the cost of hiring officers as a drain on the bottom line of the project. However, when weighed against the cost of one major theft, security officers are a bargain. They are discussed in greater detail in Chapter 7.

METALS

Theft of metals is not a new phenomenon. I came on the Los Angeles Sheriff's Department in 1976, and at that time, the metal thefts detail was in full business in Detective Division. Copper and aluminum come to mind as the most frequently stolen metals back then. Criminals would steal the steel metal guard rails from the side of curvy mountain roads just to make a buck.

Nearly 40 years later, the intensity of metal theft seems to have increased. Batteries are stolen off construction vehicles and sold for the lead inside. Battery cables are stolen along with the batteries and sold for the copper. In fact, nearly every copper cable that is exposed on a piece of construction equipment is subject to theft. Many of these cables cost thousands of dollars and are not easily replaced because of the uniqueness of each one.

This underscores the need for securing your project at the outset. Beyond the costs of the theft lies the cost of construction delays while waiting to purchase new batteries and cables.

THE INTERSTATE 5 PROJECT AT NORWALK: PART TWO

On an October afternoon, I received a telephone call from Jason Mattivi at the freeway construction site at the Interstate 5 freeway in Norwalk. He and his crew had just completed a new loop onramp at Norwalk Boulevard and had moved a crusher

FIGURE 2.2

The crusher. A marvelous invention that is vulnerable to cable theft vultures.

machine to the site (Figure 2.2). This machine is a marvelous invention because it is used to take the old concrete they had just broken up during the clearing of the old site's materials and crush it back into gravel to use in the concrete for the construction of the new segment of the project. It is a huge machine and would not be easy to dismantle and haul away.

Jason was adamant that we have frequent patrols of that machine during the night. Why? Because even though this machine would take a long time to steal, numerous expensive cables were attached to the device, and theft of these cables would delay or stop production altogether until the cables could be purchased and replaced.

These cables are not something that you can send a worker to the auto parts store to buy and be back in operation in a couple of hours. They have to be specially purchased and cost a *lot* of money. (We had been previously called to protect a crusher at another company's site because someone had stolen a long cable to the machine. That crusher cable cost $30,000.)

The point of this illustration is that it is imperative to properly secure all machinery that is left out overnight. The $30,000 cable was a high-cost example, but your job can come to a screeching halt if a thief steals the cables off of critical machinery, forcing you to locate and replace the stolen items. Again, by thinking of the costly down time, the proper security program, set in place at the outset of your project, represents a significant return on investment.

MATERIALS

Theft of materials beyond metals can put a dent in the bottom line. Whereas theft of sand and gravel is generally rare, theft of lumber from construction sites can cost a pretty penny. Both should be secured because anything that costs money can and will be stolen.

By necessity, most materials must be placed at the site near where things are being built. This is yet another reason that construction sites need a **convergence** of technology consisting of physical barriers and security officer coverage.

For example, many companies use cameras as a forensic tool to view construction sites. Cameras at night need adequate lighting to be effective. Chain link fencing should be in place around where your materials are placed, keeping in mind that you must set up your fencing so as to be convenient for loaders to access the material. Furthermore, you should have a security officer on duty to either deter thieves or to call law enforcement during the commission of a theft.

CHAPTER POINTS BY DISCIPLINE

1. **Security professionals:** By knowing the concepts in this chapter, you can advise your construction clients what to expect in the way of theft and the steps they can take to prevent or at least mitigate it. Part of your service goals should be to add value above just providing guards and equipment at a construction site. You can suggest some of the solutions in this chapter or offer to do them for the client to allow him or her more time to build the project.

   Depending on the scope of your security business, you may recommend equipment vendors or even consider adding a vendor component to your operation.

   Regardless of how the equipment arrives on site, you must make sure that your security officers know how to use all of the tools to their maximum effect and benefit.

   Furthermore, it is to your company's advantage to keep track of local construction crime trends. Reach out to the local law enforcement agency, even if difficult at first, to establish a relationship with someone in that department who is willing to keep you apprised of theft patterns in the area of your construction project. When this individual realizes that your intentions are to help with crime prevention, you might find him or her to be most helpful. Be sure to make sure that the lines of communication go both ways and offer him or her information about unusual circumstances that your security officers might observe. A small bit of information just might be the last piece to a crime puzzle that a detective is trying to solve.

   Remember that your goal must be to actually provide beneficial service to your client rather than trying to impress him or her with information that you have learned through research. You must know how to apply the knowledge and then go about noiselessly applying it to your project. Your client will notice a difference.

2. **Construction professionals:** The principles and information in this chapter will assist you in building your project with minimal worry of the interruptions caused by construction theft. You must decide before you begin your project

whether you are going to take on the additional task of protecting your jobsite or entrust it to a security company in your area. If you already have a good working relationship with such a company, that is great. When you send the security contract out for bid, keep in mind that there are considerations besides price to remember. If you hire the right company, it can save you countless hours of needless frustration chasing after the results of poor security service.

The need and duty to protect your vehicles and equipment should be on your priority list, just under getting your project built. If you are not already taking the steps to protect your vital equipment, then think for a bit about the protective measures in this chapter. If you take the time at the outset of your project to set up a proper security program, I guarantee you that you will notice a difference in the pace that your project goes.

Reflect on the protective equipment that is available. In relation to the costs associated with thefts, the cost of this equipment is minimal. Remember that even a small theft, such as the theft of copper cables from a large machine, can delay a project and cost you money. While you are negotiating your security contract, see if your security contractor already has some of this equipment. Frequently, they have it sitting in the office unused and will throw it into the deal at low or no cost.

Also, if you haven't done so, find a central place to store materials, particularly metal products. Even heavy materials, such as I-beams, steel grates, rebar, and stacks of lumber, can be removed in minutes by thieves who have big trucks and are dressed like construction workers. Remember, all they have to do is to look like they know what they are doing and like they belong at the site, and law enforcement will drive by all day and all night. You need someone on site 24 hours a day, 7 days per week who knows who and what belongs to the project site. That is the most reliable way to protect your livelihood and the integrity of your company.

3. **Law enforcement professionals:** Both detectives and patrol personnel in the areas where construction projects are underway should make themselves aware of the prevention concepts in this chapter. Face it, in most cases, law enforcement is reactive in nature. I am not a big proponent of "buzzword" concepts in police work, but proactivity is the exception to that mindset. Those of you who are shaking the trees for crooks know what I am talking about.

If you are out and about and you come upon a construction site in your area, take the time to stop and meet the foreman of the project. He or she may not have a lot of time, but usually he or she will carve out a few minutes to speak with you. Try to find out what is being built and how long the project is estimated to take. Ask the foreman what types of equipment and materials will be on site and what kind of security and security devices will be on site. Then make sure to share that information with your fellow officers so that they, too, can keep an eye on things there.

After the project is under way, if there are security officers on site, make a concerted effort to meet them. If you have never met any of these amazing people who watch things at all hours for little money and zero glory, you may be surprised. They take their mandate to "observe and report" seriously. Often they observe a lot but have no one to whom they can report what they have seen. Most of them are pro–law enforcement and are most happy to share what they see with you. I suggest that you give them your business card and a way to contact you if they see something or someone suspicious. It may sound like a pain in the neck at first glance, but when you think of what you go through to develop informants, this is easy.

Familiarize yourself with what electronic and mechanical devices are on site to deter theft and record activity. Some of this equipment may be useful when conducting a theft investigation, especially if you are on a department that requires you to conduct your own investigations instead of turning them over to a detective.

Find out what material will be on the site and who should have access to it. Keep in mind that whatever is not permanently affixed to the project can, and most likely will, be stolen. Take pictures of the equipment that belongs to the company as well as the rental equipment they may be using. Be cognizant that there are often subcontractors on site and try to find out who they might be. You wouldn't want to detain someone just because he or she has a different-colored truck than the contractor. Any subcontractor should be able to tell you without hesitation who the main contracting company is and what part of the operation he or she is working on.

Locate and identify any metal recyclers in your area. Many crooks are too lazy to take their stolen goods very far, and if you are able to notify recyclers of a theft, they may be reluctant to buy identifiable stolen metals. This won't necessarily stop a thief, but it will delay him or her long enough for you to coordinate with neighboring agencies to be on the lookout for the thieves in their areas. Better to share the credit for an arrest than to let the thieves get away.

4. **Insurance professionals:** Familiarize yourself with the precepts of this chapter before quoting the rate for the next outdoor construction site. The object of your business approach should be to maximize your profit by not having to pay claims. It is not disingenuous to ensure that your client makes few or no claims by setting the standard for his or her success. The best claim is the one that is never made. Your company makes its money, and your client makes his or hers, by being able to build projects in a timely fashion, free from the worry of crime.

Set the stage for their success by walking the site with them before construction, looking for ways to secure their equipment and materiel. Suggest that they hire a consultant if they have little or no expertise in security.

Suggest that they use a convergence of security equipment and live personnel to secure the site. Let them know that a proper security plan combined with the proper amount of insurance will help them prevent or greatly mitigate loss to a large theft. Remind them that even the theft of copper cables from machinery or the theft of fuel, tires, and batteries can cost them early completion bonus monies.

Offer reduced rates for properly secured construction sites. You may make less in premiums, but that will be offset by the payment of no or minor claims. Provide them with literature (e.g., this book) to help them with their security efforts.

5. **Legal professionals:** Depending on which role you play in the legal system, whether an attorney, a judge, or a legislator, knowing the concepts in this chapter can assist you in charging, adjudicating, or providing legislative assistance in prosecuting and punishing these criminals.

If the object of your prosecution is the maximization of punishment to serve as a deterrent to other criminals in the area, then each and every one of these cases should receive the maximum number of charges from the outset even if later you have to bargain away some of them in exchange for not having to go to trial.

If you and your colleagues are working together on these types of prosecutions, get together either formally or informally and look at the law concerning theft from construction sites. In some jurisdictions, these crimes are often charged on the nature of the items stolen, but there is no specific crime of theft from a construction site. Discuss with your colleagues (preferably with some judges in the group as well) what you believe the law in your jurisdiction may be lacking. Perhaps new legislation or sentence enhancements to theft laws are needed for crimes that take place at a construction site.

After you have looked at the law as written and have come up with solutions, find a lawmaker in your area who is receptive to writing this type of legislation. Many legislators are looking for new bills to sponsor to enhance their "tough-on-crime" portfolio, and you might even find yourself writing the new law that your lawmaker carries.

Rolling the Dice

3

CHAPTER OUTLINE

This chapter, through anecdotes and statistics, identifies the underlying attitudinal causes that lead to construction site theft. It illustrates how construction professionals as well as the clients who write Requests for Proposals (RFPs) ignore the need for security at large outdoor capital projects until a theft occurs. Therefore, the cost of security is generally left out of many proposals. Consequently, security protection causes an impact on the budget of nearly every project.

THE BOTTOM LINE

Every industry, including construction and security, is in business to make money. Without that incentive, few people would put up with the headaches of running and owning an enterprise. Everyone who works has a bottom line. Without that, there would be no reason for any of your workers to show up at the job.

As an owner or manager, you know that money is the life blood of business, and it is your responsibility to keep that blood supply well stocked. So, from the time you write the proposal for the client, you are always balancing the projected profit with the need to bid low enough to get the job. Every penny, it seems, must be maximized so you don't end up building your project for minimum profit.

You are so busy calculating material and labor costs, as well as the completion schedule, that often you forget to ask yourself how much it will cost if the project is delayed because of a theft of equipment or material or because of serious sabotage or vandalism to the project. In other words, you are "**rolling the dice**," gambling that nothing will go wrong. If nothing goes wrong, then you have won, but if you suffer a major loss, you will feel like the guy who drove to Las Vegas in a Cadillac and rode home on a bus because he lost everything. Chances are that if you suffer a significant loss, you will end up hiring security officers anyway. Why not avoid the pain and set up security in advance?

As the general manager of a statewide security company, I am always amazed that a majority of our construction jobs come after a major theft has occurred. Of course, I am always glad to provide the needed security coverage, but I ask myself why we were not called at the beginning of a project, during the planning phase. At that critical point, even if our security officers were not yet in place, I could have planned in advance for the strategies and equipment needed to properly secure the site.

You might ask yourself why this is important and why would you want to start spending money on a project when nothing has yet been stolen. Good Lord. After all, the bottom line must be watched.

I am reminded of a new client who later became a steady client. He called me because a rock crusher at one of his yards in Long Beach had suffered the theft of a major cable. As mentioned in Chapter 2, his cable cost him $30,000, and of course, it's not the kind of cable that you can run down to the hardware store and pick up in an emergency. Needless to say, he could have secured that crusher and everything else in that yard for the better part of a year with the money lost on that one cable.

The time to start thinking about security is before the first shovel full of dirt is turned. It is to everyone's advantage to consider security needs at the beginning of each project. The client wants the project built in a timely manner. The construction professionals want to complete the project on time or early so they are ready to move onto the next project and can avail themselves of any early-completion bonus. The security professionals want an early start so that in addition to satisfying their bottom line, they can develop strategies to keep the project running. Insurance professionals want to make sure that the project runs with minimal or no losses so that their bottom line suffers little.

HAVE A SECURITY PLANNING MEETING

One suggestion is to make the time before the project begins to have a meeting with the construction management team, the security management team, and a representative from the local law enforcement agency. This meeting provides the opportunity for all concerned to see the scope of the project, learn the timetable, and make suggestions for the security plan. Allow all concerned to walk the project, ask questions, and make suggestions. Discuss the times and dates when significant milestones to the project will be met.

Discuss communications, both with security and law enforcement. Make sure that all personnel know how to contact the right person at the law enforcement agency to ensure a rapid response when needed. Discuss asking law enforcement to make regular patrol checks of the site. A cup of coffee or tea or a cold beverage during hot weather will help to "break the ice" with your local law enforcement officer and will help develop relationships between construction and security personnel and the police.

Discuss the situations which might require additional security officers so that that need is preplanned for and that the officers are scheduled. The better your plan, the better your chances of success will be.

DO-IT-YOURSELF SECURITY

Some efforts to conduct security efforts "in house" can be effective if blended with a solid security plan. For instance, the installation of light towers to illuminate the site at night is a good idea because it can increase the visibility of equipment and materials for law enforcement and your security officers. Proper lighting can also assist you with the visibility for any camera system that you might have installed.

Cameras can be extremely useful as a forensic tool after a theft or vandalism has taken place. However, they do little to deter crime unless someone is monitoring them in the moment. There are no statistical data available showing that the presence of cameras will keep a criminal away.

Another word about cameras: Some construction companies and security guard companies are tempted to erect "dummy" cameras as a deterrent to theft. As I mentioned, cameras do little to deter crime, and a dummy camera can, in some cases, lead to civil liability. Your workers and the general public perceive that someone is either watching the camera or recording the events that the camera sees. If a negative event occurs, the "good guys" at the event may reasonably believe that the event had been captured on camera and that there is a photographic record of what happened. Imagine your attorney having to appear in court and explain to a judge or jury that no record of a litigious event exists because your camera was fake. This may or may not cause judgment to be rendered against you, but why take the chance? Simple solution: *no dummy cameras*. If an area needs to be covered by a camera, cut loose some cash and buy a real camera that is attached to a real recording device.

All security devices should be used in **convergence** with live security personnel.

Lastly, some construction companies are tempted to use their own construction personnel as de facto security personnel, especially if there is a site where workers are spending the night in trailers or shacks. Ask yourself if you were short of heavy equipment operators, would you draft a security officer to drive your machines? If the answer is "hell no," then don't make the mistake of using construction workers to man security posts.

Think about it. The chances are that your labor force has worked all day and are looking forward to a well-deserved rest at the end of the day. Security work can be tedious, especially during the middle of the night. Do you want your equipment and materiel to be guarded by someone who is likely to fall asleep? Do you want your stuff to be guarded by someone who has waited all day for an adult beverage and a comfortable spot to rest?

Security officers in most jurisdictions are required by law to have training in their field. All reputable security companies provide continuous professional training for their security officers. If there is site-specific training for your project, your security company should provide it for their officers as long as you make them aware of it.

In many cases, the cost of a security officer is less than that of a construction worker. Spend the money at the outset and save yourself grief later.

THE PROPOSAL

When preparing a proposal for a job, ask the client in advance if there is a provision in the RFP for funds for security if you don't see one. If you have a good working relationship with the client, you should be able to show how security can provide a positive return on investment (ROI).

Part of the intent of this book is to change the mindset in the construction industry to recognize that security costs should be an integral part of the RFP. Security guard coverage is cheap insurance and provides the optimum "bang for your buck," because live people are on site making an effective effort to deter theft and vandalism, thus minimizing loss.

An effort should be made to include sufficient funding in the RFP to hire a reputable security guard company. A reliable security vendor is essential to the success of the program, and every bidder should be either known to the builder, or vetted thoroughly prior to hiring. This means asking for references and checking them by calling past clients. This also means taking some of your valuable time to meet with your potential security provider and asking pointed questions in advance. Also, take the time to add your specific requests into the security contract. This will pay off in spades.

Lastly, it may be tempting to go with the lowest bidder in order to enhance the bottom line. If the lowest bidder meets all of your requirements, by all means, they should get the contract. You should maximize profit wherever you can.

However, if your lowest bidder is "so-so," and a higher bidder measures up to your requirements, go with the vendor who will do the better job. No security guard company can guarantee that no theft at all will occur, but the better company will do a better job over all at deterring or mitigating a theft event. That difference will offset the difference in rates, and give you a better return on investment.

CHAPTER POINTS BY DISCIPLINE

1. **Security Professionals** – As the security expert, your job begins during the proposal process. It is your job to set the tone in the minds of the construction professionals as to the importance of proper security. This goes beyond getting the contract (which is important) to developing a relationship and exposing the light of value in your approach to the security plan for the site. You need to know the costs and consequences of construction theft, so that you can overcome the objections that will inevitably arise during the sales portion of the process.

2. **Construction Professionals** – Don't roll the dice. You are the only person who can make the decision whether or not to gamble with the chances of theft and/or vandalism to your project. Consider how much a theft incident might cost, and weigh that cost against the cost of a security program. You will likely conclude that the security plan will pay off in spades.

    Make sure to include the cost of security in your bid, and be prepared to explain to your client the value of security versus the costly and time consuming results of a major theft event.

    Make sure that you and your security contractor have a good working relationship with local law enforcement. Just as you are in business to build things, they are in business to catch criminals, and once they understand the challenges that you face from crime, they will devote as much time as their schedules allow to assist you and your security team in preventing crime or catching the criminals who have stolen from your project.

3. **Law Enforcement Professionals** – As your duties permit, make an effort to reach out to the construction professionals in your area who are building outdoor capital projects. Assisting construction firms in preventing or mitigating theft should be an integral part of your community outreach program. You may or may not frequently have construction projects in your jurisdiction, but if you keep aware of the projects in your area, you can go a long way to keep the crime statistics down.

    Perhaps the construction company has not entertained the idea of holding a security planning meeting. It is entirely appropriate for you, as law enforcement professionals, to suggest such a meeting. You could suggest that in addition to a detective attending the meeting, one or more of your patrol officers who work in the area be there. They are most likely the first officers who will conduct the preliminary theft investigation.

4. **Insurance Professionals** – Whether you are in claims, or whether you are setting policy or determining rates, as an insurance professional, your goal is to save your company money and protect your client by being proactive. You can help your client and save your company money by discouraging them from "rolling the dice" with security. It is just good business to do so.

5. **Legal Professionals** – Whether you are counsel for the building contractor or involved in criminal or tort law in cases involving construction sites, you should be aware of the need for construction professionals to properly secure their sites. Often, the legal profession drives important changes in a variety of industries. Convince your clients of the importance of crime prevention as opposed to "rolling the dice."

Naming the Losses

4

CHAPTER OUTLINE

This chapter identifies almost every potential loss at any site and discusses the impact of these losses not only on the bottom line, but also on the construction timeline. The chapter uses some anecdotal examples, including photographs of everything from theft, to arson, to accidental damage, to trespass incidents, to illustrate graphically the risks.

EXPECT THE UNEXPECTED

Norwalk, California, early in the year: I received a call one morning from Jason Mattivi of Security Paving telling me that during the night, someone had come up on the freeway and torched (set fire to) a large front loader and pretty much turned it into a paperweight (Figure 4.1).

FIGURE 4.1

A torched front loader, a victim of arson.

I met Account Manager Paul Scauzillo (Los Angeles Sheriff's Department (LASD), retired), and off we went to the site. This particular project was a freeway widening, but the yard and most of the equipment was parked on the street level. Because this portion of the project was nearing completion, the front loader was parked on the freeway. Access could be made both from the freeway, if you knew in advance where to exit, and from the street below. Sure enough, when we arrived, Paul and I saw that the entire cab and much of the engine compartment had been burned (Figure 4.2). Some of the plastic material had melted in the fire and fused with one of the back tires (Figure 4.3).

There was a plastic jug, of the kind used to hide antifreeze (probably the container for the accelerant), turned upside down on a piece of rebar where the freeway side wall was to be built, and a jacket, possibly belonging to the perpetrator, hanging on another piece of rebar (Figure 4.4).

FIGURE 4.2

Burnt cab and engine compartment.

As we conducted our investigation, we spoke with a California Highway patrol officer and representatives from the Downey Fire Department and the Los Angeles County Fire Department. We learned that the fire to the front loader was the second of three fires set by a serial arsonist who had set fire to a building in Downey, the front loader in Norwalk, and an elementary school in Norwalk.

At the time of this arson, we had set up a continuous patrol, 2 miles back and forth on both sides of the freeway, along surface streets alternately with visits to the elevated roadway of the freeway. This patrol had been pretty successful up to this incident. Because the patrol route took our officer through city streets, the entire circuit took between 30 and 45 minutes to make. When our officer was back on the surface street, it was impossible to see the equipment on the freeway at night, other than a shadowy view of the top of each piece of equipment. The arsonist's fire burnt

FIGURE 4.3

Melted plastic fused with tire rubber.

the equipment and was put out by Downey and L.A. County Fire while the security officer was patrolling on surface streets, and the officer was unaware that there had been a fire until the sun had come up and the damage could clearly be seen.

We expected thieves and taggers (mindless vandals who amuse themselves by spray painting their gang monikers and other writing on walls, equipment, materials, and any other surface that will bear spray paint). We expected homeless people walking through the site looking for scrap material to steal. We expected just about anything except a serial arsonist who randomly picked our front loader to burn.

This is why, when securing outdoor capital projects, you have to expect the unexpected and prepare as best as possible for every event. This incident serves to illustrate that in some cases, it is wise to use two vehicles and two security officers despite the cost. I will explain further in Chapter 7.

FIGURE 4.4

Possible evidence, a container and an abandoned jacket.

WHO ARE THE THIEVES?

What does a thief look like? What does a criminal look like? Any experienced cop in the world will tell you that although some street people by their appearance cry out to be stopped and investigated, most of the time, you can't tell by just looking.

The fact is that in business, at least some percentage of company theft is perpetrated by employees—the people who you "let into your house" when you hired them. In my CPP studies, I learned that a majority of large thefts from organizations are perpetrated by people who enjoy elevated positions of leadership (i.e., top management).

Does this mean that you have to be paranoid about those who work for you? Perhaps. Perhaps not. Although I believe that almost no one takes a job with the intent to steal from their employer, some employees in time devolve into thieves.

So the question becomes, what can you do to deter employee theft, save yourself money and grief, and save your top management from prison?

THE BACKGROUND INVESTIGATION

To start with, take the time to do a thorough background investigation at the beginning of the hiring process. Don't assume because people have cordial personalities and appear to have the requisite skills that they are who they say they are. Ronald Reagan used to say, "Trust but verify," and that should apply in your hiring, even if you really like or think you know the applicant. Being thorough in checking backgrounds is simply good business.

Find yourself a reliable local company to do the criminal background check. Each local jurisdiction has different laws regarding applicant privacy. For instance, in California, you can only go back 7 years of criminal history when considering applicants. I won't go into my theories of why that might be the law, but you can see that someone's criminal history is important to your hiring decision.

If an applicant gives you references, don't just nod your head and feel impressed by them. Pick up the telephone and call the references. When you talk to the references, ask her person if he or she can suggest anyone else to call. You might actually find out about an employer who is not listed in the employment history. When you call that employer, you might find out why he or she is not listed.

Always ask your applicant to provide you with a driver's history report from your local jurisdiction's Department of Motor Vehicles even if the employee will not be operating a vehicle or piece of equipment. Look at the report for records of driving while intoxicated. A conviction or a series of convictions for driving under the influence will tell you whether you need to be concerned about substance abuse problems. This goes beyond worry about whether the employee will operate equipment or vehicles while intoxicated. It speaks to the associated risks with substance abuse. Will this employee end up stealing from you because he or she needs money to support a substance abuse habit?

Does this mean that anyone who has ever had a substance abuse problem should never be hired? No, but what it means is that you need to ask the applicant to provide you with documentation that he or she has attended and completed a legitimate substance abuse treatment program.

If your local jurisdiction allows it, take a look at the applicant's credit report. I know that in today's world it, has become fashionable for average people to run up a mountain of debt to buy the things that they believe they need to survive. You will most likely never be able to change that. However, you can see from the credit report if the applicant is "snowed under" with debt and whether he or she has a record of paying his or her bills and paying them on time. Excessive debt can create tremendous pressure, which might cause an employee to rationalize in his or her own mind that theft from the employer is the only way out of the situation. Later I will discuss the importance of steering your employees to programs that teach them how to properly manage money.

Another character trait that you must look for in your investigation is the propensity of your applicant to gamble. This is harder to ascertain. It is difficult to tell from your employee interview whether or not the applicant has a gambling problem because similar to most addicts, gamblers won't admit that they have a problem. You may get lucky with a question such as, "What is the most money you have ever lost when gambling?" The question may catch them off guard and lead to a discussion about gambling, but then again, it might not. During your background investigation, you are more likely to get an honest answer from someone who knows the applicant. It is perfectly legitimate to ask the gambling question of a reference. Most people will be truthful about another person's foibles, but they may not be honest about their own.

The importance of a gambling problem cannot be understated. Depending on the type of gambling they do, they could already be involved in a shady, pressurized world. If someone makes an occasional visit to a casino, that doesn't make them a problem gambler so be judicious when considering someone's gambling habits.

The background investigation is also detailed in Chapter 9 under the heading Employee Screening.

THE THEFT TRIANGLE

ASIS's Protection of Assets manual describes three elements that are always present when employees steal. The three corners of the triangle are **motive**, d**esire**, and **opportunity**.

The late sociologist Donald Cressey broke **motive** down even more specifically when he described a "perceived un-shareable financial need" in the mind of the thief. After all, no one wants others to know about financial, substance abuse, or gambling problems.

Sometimes the motive can be to satisfy an imagined wrong on the part of the employer, such as being passed over for a raise or a promotion, or some other perceived maltreatment. Whatever the justification in the thief's mind, be aware that it is just an excuse. The motive is something that you cannot control.

Desire is created in the mind by the thief and is generally related to the motive in some way. The desire can be as simple as wishing to bring a financial problem to an end. It can be to exact revenge on the employer for a perceived wrong. It can also be simple greed, wishing to improve their lot in life by stealing from someone who has more than they do. Desire is something that you cannot control.

The last corner of the triangle is **opportunity**. None of the other two corners of the triangle can come into play if the thief does not have the opportunity to steal. Through proper use of deterrents, controls, audits, and physical security, you can greatly decrease the opportunity for theft.

Opportunity is something that you *can* control. In fact, it is the only part of the triangle that depends on you. The security concepts in this book must be applied internally as well as externally and are addressed later as I discuss strategies for theft deterrence.

THEFT BY OUTSIDERS

Traditionally, we think of thieves from the construction sites as outsiders, and in many cases, they are. That is why a comprehensive plan for security must be in place.

Thieves appear to be work averse, but that is not so. They observe, plan, and execute. Most thefts require thought and effort. Some thieves put more effort into their crimes than some working people put into honest work. Many of them operate under the belief that they will never get caught, or if they are caught, they will get away with it. In many cases, they are right, which is why you must be prepared and vigilant to stand up against thieves.

Keep in mind that outsider thieves use the same **theft triangle** that insiders use and that **opportunity is the only aspect of theft that you can control**.

Outsider thieves range from homeless people who wander about your site looking for scrap metal or lumber to pick up to sophisticated thieves who plan to drive off with the most expensive piece of equipment that you own. The best deterrent to these thieves is a strong convergence of security personnel, good lighting, and good security equipment.

EMBEZZLEMENT

Embezzlement is the theft of property by a person to whom it was entrusted. Most of us think of embezzlement as theft of money by clever by unscrupulous office types, and that is certainly part of it, but it also applies to equipment and materials that are stolen by employees.

To be sure, no business owner wants to think that the people who they hired and to whom they turned over the keys to the kingdom have stolen from them. This kind of theft goes beyond the simple loss of money and materiel because it is such a betrayal of trust.

The embezzlers can be the friendly people who greet you every day when you walk into the site office or the quiet ones back at the main office who go about the daily grind of crunching numbers, issuing paychecks, and paying bills. Money is the life blood of business, and those who handle it enjoy a tremendous amount of autonomy because most people don't want to be bothered with the minutia associated with accounting and payroll.

So what do you do to deter or prevent embezzlement? The answer is the same as in any other type of theft: **Control the opportunity**.

When you are setting up an operation, make a visit to an independent accountant and prearrange regular but unscheduled audits of the financial component of your operation. Embezzlers, similar to other thieves, don't want to get caught, and the expectation of an audit should deter the faint of heart. There will still be the one or two embezzlers who think they can beat an audit, but most don't want to take the chance.

Make sure that all checks, both for expenditures and payroll, require a double signature. One of those signatures should be a company owner, general manager, or other trusted employee who is outside of the payroll and accounting office. Make sure that the trusted signatory is not a friend or relative of the accounting person.

If the accounting is still on paper, make a habit of looking at the books regularly. Ask questions about anything that looks amiss and weigh the answers to those questions against logic.

If the accounting is electronic, you can and should check on it various times of the day or night. If accounting entries are time stamped by the computer, look for entries that are made outside of normal working hours. If you have a money-handling employee who regularly works outside of normal hours and who never seems to take vacation or holiday time off, there is at least a possibility that he or she might be embezzling.

There is an old saying that says "just because you are paranoid, that doesn't mean someone isn't following you." Be a little paranoid about your money. It will pay off in spades.

The other kind of embezzler is the trusted employee who is adding a room to his house with your equipment and material. In some cases, this is literally true, but the theme is the same whether the person is stealing company tools, equipment, supplies, or materials and either converting it to his or her own use or selling or trading it. You must take the time to develop a control system for everything at the site, from paper clips to steel girders to machinery. Try to think of everything on site as a pile of cash. Some piles are small, and some are large, but they all add up. Ask yourself, if you saw 50 cents or an $100,000 lying on the ground, would you leave it there? I don't think so. Then why would you allow cash in the form of company property to walk off site?

Control systems for both money and equipment involve monetary cost and take time. You have to pay someone to attend to these matters, and you have to take the time to listen to the person, heed his or her advice, and make changes if necessary. Look at these things as an investment. If you prevent even one major theft, your return on investment will be high.

THEFT OF MACHINERY

Back to the Security Paving Norwalk site. Several months after the arson to the front loader, I got another call from Jason Mattivi. This time a backhoe had been stolen during the night while one of our security officers was on duty (Figure 4.5). I will not tell you what was going on in my mind as he was telling me this. Some young person might pick up this book.

As a security manager, I was asking myself where was the officer, and what was he doing? Was he asleep? Had he left the post?

Again, this site stretches for more than 2 miles along the Interstate 5 Golden State Freeway on both sides. The backhoe had been parked across the street from the site yard and had been hooked up to a trailer and driven away.

The fact is that the officer was in fact on duty, driving his patrol vehicle dutifully around the site. The thieves had been apparently watching his activities and had

FIGURE 4.5

How fast can a backhoe be stolen?

waited until he left for his patrol. As soon as he was gone, they grabbed the backhoe, put it on a trailer, and drove it away.

Luckily, this story has a happy ending. (Many of these stories do not.) Security Paving invests in LoJack tracking devices for all of their equipment. They activated the device and found the backhoe in a neighborhood about 7 miles from the construction site and recovered it. I will discuss LoJack and other devices in the Chapter 8. I will also later discuss a security strategy that might have reduced the chances of this kind of theft.

THEFT OF MATERIALS

When one thinks of theft of materials a construction site, the theft of copper almost always comes first to mind. Beside the giant spools of copper-filled cable that often can be found at sites, copper can be stolen off about every single machine on site.

However, lots of other materials are targets for theft. Keeping in mind that one bundle of rebar costs around $60,000, it's not hard to imagine the likelihood of theft of this type of material for resale to other construction companies or even back to you.

Sadly, some of this material is set up for theft by workers at the site. I can recall some rebar grates (weighing about 150lb apiece) at a freeway construction site that had been moved from where they had been set down and propped up on a wall adjacent to a road, obviously having been set up for theft. It had been raining off and on before the incident, and the ground was soft enough to allow footprints to be clearly seen. The footprints showed that no one had "hopped the fence" but rather that the grates had been moved from their original location by someone on site for later theft.

Stacks of lumber are also targets for theft. Again, don't place your confidence in the fact that a law enforcement officer will dive by during a theft and stop it. As I have said, most officers will think that a move of lumber is legitimate if "workers" in construction clothes wearing hard hats use a forklift to load stacks of wood onto a truck or trailer.

Thefts of all of these kinds of materials can be most effectively mitigated if there is someone on site who knows what materials are on site and who should be moving them and at what time. Creating a secure fenced-in storage area coupled with a time-table for deployment of materials is a start to controlling material theft. Someone, either a construction professional or a security officer, should be assigned to the materials yard and should document in writing every move, detailing who received the material and to where in the project it was destined.

At the end of the work day, a worker should go to every place at the site where material is left and exposed and take digital photos to pass on to the night security staff so they will know the areas to which they should concentrate their patrols.

I'm sure someone is saying, "But that will cost a lot of money and will take a lot of time." That's quite true. It will cost some money, and it will take some time and effort. However, when you weigh those costs against the theft of say, $ 60,000 worth of rebar, you can see where that small investment of time and money could produce a huge return in reduction of theft. This type of program can be built into the security portion of your proposal, transferring the cost to the client.

FIRE

Back to the bridge in Hesperia. One Monday afternoon early in May, I got a call from Pacific Protection Services Field Supervisor Travis Jones to let me know that the Ranchero Road bridge construction site had caught fire. This bridge is eight lanes wide and was designed to connect the east portion of Hesperia to the west side and give the residents easier access to both sides of the city.

At the time Travis called me, the steel girders and rebar had been set in place, and the plywood framing for the concrete was in place. This signaled that in a matter of a couple of months, the project would be finished, and we would move on to the next one with our clients. In my mind, having seen the completed concrete ramps already in place, I wondered how serious this fire really was. After all, concrete and steel don't burn, do they? However, knowing Travis Jones to be a serious worker and a good supervisor, I knew he wouldn't call over a minor incident.

I asked him what had happened and how was it impacting the project and our job as the security team due to come on that night. He told me that the highway patrol had closed the freeway in both directions and was routing traffic around on side roads, thus delaying motorists for hours. Keeping in mind that because so many millions of people in Southern California travel the freeways every day and night of the week, a complete shutdown is usually only done for a fatal crash on both sides of the freeway, a multi-car pile-up, a plane crash, an earthquake, or some other serious disaster, I knew this was a serious situation (Figure 4.6).

FIGURE 4.6

Bridge fire in Hesperia. *Photo Credit: KABC7.*

When I went to the television screens in the command center and saw the news coverage, it was hard to comprehend that this bridge was the same one that we had been securing since construction began. The steel I-beams that stretched across the span were beginning to bend into a snakelike shape, and it was soon apparent that the whole bridge, as constructed so far, was destroyed. Debris was dropping onto the freeway, and nothing could be done until the fire burned itself out or was finally extinguished by the magnificently brave and tireless fire crew.

The highway patrol gave Security Paving 24 hours after the fire had been completely extinguished to demolish what was still left standing and clear the roadway for traffic.

The investigation revealed that a contract welder had been welding some rebar over where some of the plywood forms had been placed, and they caught fire from the sparks of his torch. That simple act of carelessness delayed the completion of the project from July of that year into February of the following year.

Whys is this story important to securing outdoor capital projects? It serves to restate the admonition to expect the unexpected. Be prepared to put in place your disaster contingency plan for that site.

INJURY ACCIDENTS

Many contractors look for security officers with first aid training. Some security companies require it, but not all. This is hard to comprehend, particularly when providing security at large outdoor capital construction sites.

If one looks at a security plan as more than just protecting "stuff" but rather protecting assets, then the goal of preventing accidents and applying first aid if an accident occurs makes good sense. Most corporate executives worldwide would probably agree that the people who work for their organizations represent the most valuable asset that they have.

The convergence of security officers with technology should be involved in the noncriminal aspects of security as well as theft prevention. That means that part of the security plan is to have first aid trained officers who are equipped with first aid equipment and other lifesaving devices and who are trained in the use of those devices, (e.g., defibrillators, Ambu Bags, EpiPens, splints, stretchers). Officers should be trained to stabilize accident victims and then use communication equipment to summon help.

FATALITIES

Sadly, people die every day for a wide variety of reasons. However, when someone dies at work, a plethora of considerations come into play. A balance must be struck between the needs of the deceased and his or her family and the protection of the company and its assets. A properly trained security force can help construction site managers during these most difficult of times.

It would seem that the first order of business would be to summon the authorities. This should be the first thing that comes to mind even if the death seems to have occurred from a preventable accident. Generally, the first call should be placed to law enforcement. Law enforcement officers are trained and experienced in these matters and have the proper connections with the coroner and other investigative agencies to begin the inquiry and to remove the body in a timely manner.

Great care should be taken not to move the body and to establish as wide a perimeter as feasible around it to assist in the investigation of the scene if the death was an industrial accident or a homicide. After it is determined that the victim is obviously dead (i.e., no pulse, no breath, fixed and dilated pupils, and other signs), no one should be allowed near the body in order to preserve any physical evidence. It is proper, in most cases, to cover the body with a blanket or tarpaulin until authorities arrive.

(Note: If there is even the remotest question that the person might be alive, then resolve the issue by summoning medical assistance immediately. In some cases, vital signs may be so suppressed that the victim appears dead when he or she is not.)

When death occurs at a jobsite, always keep in mind the feelings and perceptions of the decedent's coworkers and other personnel. Some people will be debilitated to the point where they cannot work. Others will pretend that everything is fine and wish to continue working. Regardless, make sure to keep people at work until the authorities arrive so they can be properly identified and their statements can be taken by law enforcement if it is necessary. After that, if your employees are upset, then it is advisable to send them home or to provide some sort of counseling for them, if available. Make sure that those who decide to stay and work are not displaying bravado.

These people may fall apart emotionally when the sting of the death has gone away and the enormity of it sets into their psyche.

If you decide to keep your worksite going, keep all activity, particularly dust-generating activities, away from the body. This will help preserve evidence if it is needed.

Although it may seem crass, after you have summoned the authorities and stabilized your workforce, then the next two phone calls should be placed to your corporate attorney and your insurance agent. Remember that the goal is to protect assets, and although at the beginning of the death incident the decedent's family may seem to be "on your side," a single visit to their attorney could change that relationship, and you must be prepared for the worst.

Occasionally, a death at a jobsite may be the result of a murder or manslaughter. If this appears to be the case, then prudence tells you at that time to make sure that all personnel stay at the site until law enforcement arrives. Note the names of anyone who scrambles away in a hurry and be prepared to give law enforcement their names, addresses, vehicle descriptions, and likely destinations. They may or may not be suspects, but there is no point in taking the chance.

For purposes of your own liability, take plenty of pictures to share with your attorney, your insurance company, and the police. Make sure you keep copies of all photos before you give them to law enforcement because they are not likely to give them back. In today's world of digital photography, it should be easy to email or text them to yourself, your attorney, and your insurance agent before turning them over to the government. If they seize your camera or cell phone before you do this, you may never have access to those photos again.

VEHICLE COLLISIONS

Vehicle collisions are not unique to construction sites, but they can be costly if the circumstances involve negligence or the appearance of negligence. Remember that when trial attorneys go fishing for money from the pockets of the parties involved, they look for the deepest pockets around, which will likely be yours if you are a business owner.

That is why you must document, photograph, and take witness statements any time there is a vehicle collision, no matter how minor the collision appears. Even if you are not planning on repairing your vehicle, you need to record everything in preparation of potential legal action.

Furthermore, there are two assets you must have at all times, but particularly in the case of a vehicle collision: a great attorney whom you trust implicitly and reliable insurance. The absence of either one could cost you a lot of money. Anyone who operates without these two assets is needlessly rolling the dice.

The first considerations in any vehicle collision are the life and safety of those persons involved. In the case of a major collision with significant injury, no one will have to tell you to call for emergency medical assistance and law enforcement, but in cases wherein there is a question as to the seriousness of injuries, you should always resolve that question by summoning medical assistance. If they determine that no one is hurt, then no harm has been done, but if no medical assistance is requested and you have to explain why later, it could look bad if things are litigated. If there are fatalities, then it is incumbent upon you to secure the site as much as possible, pending the arrival of law enforcement.

If possible, take your photographs before the vehicles are moved. Always keep in mind personal safety before taking your photos. Always secure the site first. The best photos in the world won't do you any good if you get squashed flatter than a tortilla by a speeding motorist. Block off the area as though you were blocking the street for some construction obstacle.

Take photos not only the vehicles themselves but also the tire tracks or skid marks behind each vehicle. These photos will help if there is a forensic traffic investigation. Also take photos on the ground underneath the vehicles after they are moved before a fastidious tow truck driver sweeps the evidence away.

Photograph every possible view of the crash scene at a distance from the point of view of both drivers for some distance as they would have been approaching the collision. Also, after identifying witnesses, photograph the scene from the place where they witnessed it to assure that they could have seen what they claimed to have seen.

Take statements from both parties of the collision while it is still fresh in their minds and before they have time to alter the narrative to fit a more profitable narrative. Often, after the parties have spoken to other witnesses, they begin to "remember" things that they could not have possible experienced, and shortly those false memories will become fact in their minds. If you can make a video or sound recording of their statements, that is even better. That way, if stories change, juries can see what was said at the time of the collision.

Also, make sure to take statements from any witnesses who come forward or whom you find after seeking them. Eyewitnesses are great, but even "ear witnesses" can be useful for establishing the time of a crash. Make sure to tactfully ask them if they actually saw the crash, heard the crash only, or are just repeating what another witness told them.

This mini-investigation of your own could greatly assist the legal and insurance professionals who are assisting you. The subsequent law enforcement investigation will most likely be thorough, but it is possible that even the best traffic investigator could inadvertently miss a detail that will later be important.

Write a report of the incident and include your photographs. Give a copy to your attorney and to your insurance agent (who will most likely conduct his or her own investigation). Also, give a copy of the report to the investigating law enforcement officer unless your attorney advises otherwise. This also shows an openness on your part and precludes an argument in court that you were altering the facts to avoid culpability. If you do an honest investigation, the facts will speak for themselves. In some cases, your driver may be at fault, and the more honestly and clearly you approach it, the better the chances for a just and less painful resolution.

FLOODING

Those who live in rainy areas don't have to be warned of the dangers of flooding. You are well aware of the peril associated with excessive rains and flooding and most likely to take reasonable precautions every day.

However, construction security professionals the world over need to preplan for disaster when it comes to rain. Take a lead from the construction crew itself. Most site managers will stop the work with the approach of a severe storm. The catch is that security goes on even when the construction stops.

If you know about Southern California, you might think I'm daft when it comes to flood preparation because I live in an area that is just one degree above a desert. Why should those who live in arid areas be concerned about flooding?

The soil in dry areas is often quite sandy, with a shallow hardpan and very little organic matter to help absorb the moisture. Generally, the water percolates down to a hardpan, and when that water stops sinking in, Katie, bar the door. The water has to go somewhere, so off it flows. Most people have heard of flash flooding, and therein lies the problem when heavy rains come to an arid area (Figure 4.7).

Anyone who is a SCUBA diver, like I am; a boater; or even a swimmer knows the power of moving water. Just turn on the television during the reports of flooding, and you might see cars and even houses floating down a river.

Particularly in flash flooding, moving water can excavate a huge hole in a roadway and then fill the hole with water, thus making it appear like any other puddle after a rain. Then either a security patrol officer or a construction worker drives a vehicle across the puddle and sinks without warning. If the driver is lucky enough, the vehicle will only sink deep enough to embarrass him or her and can be easily

FIGURE 4.7

A security vehicle in a "shallow" puddle.

pulled out of the pit. But if the puddle is deep, it can cause the total loss of a vehicle and in extreme case lead to a drowning.

A simple admonition to any security officer of "stay on the high ground" seems like if followed, it would go a long way to preventing a long way to preventing disaster and in the case of the puddle example, it would. However, in areas where soil has been moved and not been compacted, even the high ground could bring danger of sliding.

So, what to do? No matter the climate, with today's satellite technology, most inclement weather can be anticipated. The time to mitigate the danger is before the rain, when you have a clear view of where the safe places at the site might be. Keeping in mind that the main goals of construction security are public safety and prevention of theft, the ground is not going to be more accommodating for thieves than it is for your security officers, so the goal should be safety first.

Survey the site in advance and identify the areas less likely to be compromised by floodwaters. A safe place on the side of a paved road, where one can see a majority of the site, works best. If your site does not have such a roadway, pick the highest packed soil that gives both safety and a clear view of the site. Then tell your personnel to stay at that fixed position until the storm has passed and the waters have subsided. If safe to do so, send your field supervisors to the site to make sure they are in compliance.

Many years ago, as a deputy sheriff, I patrolled the Angeles National Forest and the cities below the forest. When there was danger of flooding, I carried an inflated inner tube in my car to serve as a flotation device in case I was caught in a flood. Although I was never caught in a flood, my old partner, the late Deputy Amos Lewis, had been up in Big Tujunga Canyon. It was this simple flotation device coupled with his luck in finding a tree branch to take hold of that saved his life. His radio car was never found, but he was alive thanks to a little bit of preparation.

How much less should we do for the security officers who work for us? Flotation devices are usually available at a nominal cost at most sporting goods stores. Make sure that if your officers are in a vehicle when they are working for you that they have a way to survive.

EARTHQUAKES

Before you say that earthquakes only occur in California, South America, and Asia, think again. One need only look back to August 23, 2011, to an earthquake whose epicenter was in Louisa, Virginia, and that measured a 5.8 magnitude reading on the Richter scale. This earthquake severely damaged the National Cathedral and the Washington Monument in Washington, DC. The damage took 2 years to repair.

Earthquakes also occur with a degree of regularity in the British Isles. The Llyn Earthquake of 1984 centered in Gwynned, Wales, for instance, measured 5.4 on the

Richter scale and could be felt throughout Ireland and Western Britain. There have been many shakers in the British Isles both before and since.

Whereas most earthquakes occur in coastal regions, the fact remains that they can occur anywhere. So what do you advise your officers to do in an earthquake? Keep in mind that if the earthquake is large enough, it can cause any structure to collapse, so if your personnel are already outside in the open, advise them to remain in place unless the earth is opening up. (Then it's okay to run… really.)

You might ask, isn't it better to take cover under a concrete structure of some sort? Look back to the Sylmar, California, earthquake in 1971 and the Northridge, California, earthquake of 1992, and you will see that in both quakes, the identical segment of the freeway overpass that connects the Antelope Valley Freeway to the Golden State Freeway collapsed, causing horrible deaths to those who were crushed. Tell your people that if they are in the clear, covered only by the blue sky, then they are in the best possible place in an earthquake.

If your personnel are inside a structure, tell them not to run outside. When they run from inside a structure to the outside, they incur the greater risk of being struck by a falling object. The best place for them to take cover is under a heavy piece of furniture such as a desk or in a doorway if one is nearby. Most people who are killed inside structures are killed by falling debris. The likelihood of surviving an earthquake is good if you can position yourself so as not to get crushed.

After the ground has stopped shaking, the security officers should place themselves in a position to help others. At this point, all survivors of the quake should join together to make rescues, either at the direction of the construction site manager or the security officer if no leadership is present.

This is where training and planning become important. This is also why it is preferable to hire security officers with first aid and CPR training. When you review candidates, look at such people as volunteer firefighters, members of Civil Air Patrol, or volunteer search and rescue team members. This type of training becomes extremely valuable during natural disaster.

WATERCRAFT ACCIDENTS

Watercraft accidents are rare at construction sites unless you happen to be building a pier or if you are at a gravel mine. Gravel can be mined from the earth for such a long period of time that what started out as a dry mine ends up descending beneath the water table.

One such mine is the Hanson Aggregates gravel mine in Irwindale. The mine has been in business for more than 100 years, and a rather sizeable lake exists there; the mining is now done through dredging. There is an island in the middle of the lake, and three or four boats service the island and bring workers to and from it. Over the years, waterfowl have inhabited the lake and brought fish eggs with them that were stuck to their feet. (This has created another security problem because local residents often sneak on the mine site to fish.)

If your security officers are operating a boat at any construction site where there is water, then it is mandatory that you spend the money to send them to a government-approved boating safety class. In the United States, these classes are often conducted by the U.S. Coast Guard Auxiliary, a volunteer organization of experienced boat sailors. This type of training reduces the risk of mishap caused by mishandling of watercraft, thus reducing danger of injury or death and somewhat mitigating civil liability.

AIRCRAFT ACCIDENTS

Most likely, you will never experience an aircraft accident at a construction site unless you are building, repairing, or replacing a runway. Most airports require that you take and pass an airport driver's class and be issued a permit before being allowed to operate a vehicle on that property.

Aircraft always have the right of way over ground equipment. If you keep that in mind, you will greatly reduce the likelihood of an incident involving your personnel. In most cases, both security and construction vehicles must be escorted by a flagged official escort vehicle. This not only ensures safe passage around the runway area but also increases the likelihood that your equipment and personnel will be seen by pilots.

Just like at any other construction site, anything can and will happen on a runway construction project. When I was a boy in the early 1960s, I lived in Lompoc, California, where they were building a new airport, which is still in use today. Some hapless pilot decided to land his airplane while the concrete of the runway was still wet. Needless to say, the photo made the front page of the *Lompoc Record*, the airplane was pranged, the pilot was permanently embarrassed, and the construction crew who had poured and worked the concrete were not a little angry.

Because a majority of aircraft accidents occur during takeoff and landing, there is at least the possibility that a crash could occur while you are working. If you happen to be working at a crash site at a large airport, you will notice that a huge volume of fire and rescue equipment will respond to the incident. If you are working at a small municipal airport, there may be little or no rescue equipment, and your crew should be prepared to assist in effecting a rescue of trapped or injured people if at all possible.

When a rescue is effected, the victims should be moved away from the scene, and everyone should stand well clear of the aircraft. The reason for this is that there is a very real danger of spillage of fuel and thus a danger of fire.

Either before or during the rescue, someone (who is predesignated) should call for the fire department and law enforcement. Generally, the airport authorities, if present, will notify the Federal Aviation Agency in the United States or whatever government aviation agency that exists in other countries. Aircraft accident investigations are investigated in the United States by federal agencies, but take your own pictures anyway in case someone tries to blame you for the crash. (Remember to always think of your attorney and your insurance agent.)

CHAPTER POINTS BY DISCIPLINE

1. **Security professionals:** Expect the unexpected. The security business exists because of the unexpected things that happen in the world. By successfully formulating a security plan and by successfully executing that plan, you can help your client to be prepared when an unexpected event occurs. If you study and are prepared before each deployment to an outdoor construction site, you can protect your client's project, protect your company's reputation, and save everyone a bunch of money.

   Part of this plan should lie in training your security officers and making them aware of what to do when emergencies or catastrophes arise. That training may not only protect the client's operation but may also save lives. You can be proactive or reactive. The choice is yours.

2. **Construction professionals:** Expect the unexpected. You are out in the world creating your construction masterpieces, and your magnificent structures will be around long after everyone who is alive today on this planet passes away. Unless you have a major event such as the fire at Hesperia, there will probably never be much historic record of what happens at your construction site.

   Keeping in mind that unexpected events and catastrophes will delay your project and cost you money, one of your construction goals should be to have a relatively uneventful project. By being prepared for any kind of incident, large or small, you will maintain your company's reputation for timely completion of projects and will keep your business alive. There is an investment of time and money required for this, but your advanced attention to potential disaster will pay off in spades.

3. **Law enforcement professionals:** Expect the unexpected. That is pretty much standard operating procedure for law enforcement. When it comes to outdoor construction sites in your jurisdiction, you can become better prepared for significant law enforcement events by getting to know the construction and security professionals at the construction jobsites. With these relationships, you will know who is prepared and who isn't. With this knowledge, you can know where to expect assistance when things go seriously wrong.

   If the disaster is widespread and your construction sites are well secured, you can adjust your response priorities to other areas.

   Also, by developing these relationships, you will have your finger on the pulse of crime when you can't be everywhere at once. By talking to the security and construction professionals at these site, you can become aware of pending criminal trends and take measures to suppress crime and effect arrests.

4. **Insurance professionals:** Expect the unexpected. Every situation in this chapter could come across your desk and make its way into your case load. No one can stop criminals from plying their trade, and certainly no one can stop disaster from striking. However, by assuring that your clients are at the very least prepared for the unexpected, you can help mitigate loss.

　　If your construction clients do not have an emergency plan or a security program, you can help them by prodding them to think about potential loss and to attempt to prevent or at least mitigate the effects of serious incidents.

5. **Legal professionals:** Expect the unexpected. In the legal profession, you have the luxury of time to analyze events after they have happened and hopefully come to some reasoned and meaningful conclusion as to why they occurred and what should be done both in the present and in the time to come.

　　Through client counsel; litigation; and in some cases, legislation, you have the unique ability to impact how things are to be done in the future. It is my hope that through your efforts, standards are set throughout the world for the security industry and things improve because of your work.

Assessing the Risks and Potential Losses

5

CHAPTER OUTLINE

This chapter gives guidance to construction professionals and security managers on how to look at a construction site, new or ongoing, with "fresh eyes." It teaches them to assess the site, the area, the personnel, the equipment, and the materiel to plan for loss prevention and mitigation. It teaches them to plan and budget for the unexpected based on their site survey.

THE SITE SURVEY AND VULNERABILITY ASSESSMENT

Many CPPs and other security professionals throughout the world make outstanding site surveys every day. My hat is off, especially to those who survey construction sites before one spade of dirt has been turned. The reason for that is that they must look at construction plans, confer with the site manager, and then develop a plan to secure something that is not yet there. They have to see in their minds that some structures that are there at the beginning of the project may be demolished and therefore will no longer be landmarks. This requires a degree of abstract thought and vision. So how do they do it?

Most people don't wake up one morning with the innate ability to make site surveys and vulnerability assessments. Their craft begins with an interest in security, years of experience, and education in asset protection.

Jim Grayson, CPP, a fellow author and a friend of mine, came into the field via a circuitous route, as did I. Jim started his working career as a reserve Los Angeles deputy sheriff, moved to a regular police officer position with San Marino Police Department (California), and finished out his law enforcement career as a sergeant and later a lieutenant with the California State University Police Department.

After retirement, he entered the security industry as the security manager for the Huntington Library in San Marino, California. Through this experience, he learned quite quickly the frustration of being reactive in his approach to security. He wanted to learn how to take a proactive approach to security and develop an atmosphere where no crime existed. In other words, he wanted to make the criminals go away. He knew that this goal would require of him creativity and an education in things related to security. Years of being in law enforcement had still not given him sufficient tools to make this happen. So, he set out to gain the knowledge and skill he needed.

He learned about and joined ASIS International and became surrounded with like-minded security professionals. He studied for and earned his CPP in 1996. Even though the CPP is a comprehensive body of study, the things he learned only whetted his appetite and caused him to dig deeper and learn about such things as CPTED (Crime Prevention Through Environmental Design) and lighting.

Jim began doing site surveys and vulnerability assessments for a variety of organizations, including the company for which I now work. It was by meeting Jim Grayson and observing his work that I began to learn the basics of this critical approach to security planning.

The site survey and vulnerability assessment should be a prerequisite for every large outdoor capital project. It should be done before the project is begun and should be done regardless of whether security is from an outside source or is handled within.

Some security companies use these assessments as a sales tool, which is good but meaningless if not properly done. Construction companies should insist on these as a condition for winning a bid because the cost associated with proper planning is negligible compared with a potential loss. Many security companies do the site survey and vulnerability for free as a part of the value-added portion of their proposal.

To begin with, there is no "one-size-fits-all" vulnerability assessment for outdoor construction sites because **every site is different**. However, there is a certain commonality among these surveys, and we will touch on those.

CONSTRUCTION PLANS AND THE SITE VISIT

Before beginning your site survey, a meeting with the site manager is imperative to look at the plans for the project, determine the geographic size, and learn from the site manager what his or her timeline and concerns are in terms of vulnerability of equipment and materials. This meeting is critical. You may get only one chance because after the project has begun, you may only be able to meet with the site manager for a few minutes at a time. I assure you that from that point on, the site manager will be too busy to deal with what he or she might consider to be small matters.

If you can get a copy of the construction site plans (or at least borrow them), it will greatly help you to see where things are intended to be built and possibly to see where the main construction yard is to be placed. Take them with you to the site.

Next, you *must* visit the site. This is imperative. All the plans and photographs in the world will not help you get an accurate picture of the site as much as will visiting it in person.

When you get to the site, find a pace to safely park your vehicle, get off your rump, and get out of the car. Yes, I am recommending that you actually *walk* the site instead of sitting in your vehicle and observing from the temperature-controlled atmosphere of your jalopy. (You may have to drive to several places if the site is large enough, but you get the idea.) First of all, the exercise won't kill you, and second, you will get a much more complete feel for the area.

By walking the site, you can feel if the soil is loose under your feet, you can smell the air, and you can actually see your surroundings, and by walking, you can discover things that might be hidden from view that you would not see if seated in your vehicle.

Bring your camera and take a lot of photographs. These photos will be invaluable as you explain your conclusions in your report. They will also help to jog your memory about details that you may have forgotten by the time you get back to your computer to write your report.

Look to see where or if there is vehicular traffic around the site and be aware that you will have to plan for thieves who might have quick, easy access to vehicles, supplies, and equipment. Note the amount of traffic and the speeds at which traffic flows. This will help you suggest design improvements for access to and egress from the site.

Assess how many officers you will need in convergence to your equipment to effectively do the job. If you are building a long bridge over a freeway, expressway, river, or railroad tracks, you might consider posting officers on both sides of the project. The bridge at Hesperia (see Chapter 4) was such a project. The project stretched over a 10-lane freeway with a wide median. There was equipment on one side that you could not see from the opposite side of the freeway, and it took from 10 to 15 minutes to travel to the other side. The crooks knew this and would strike on one side of the freeway just as soon as our officers had left that side. Two officers would have mitigated that problem.

Make an alternate plan to bring a patrol vehicle to the site if the manager will not agree to a second security officer.

Determine where the site manager plans to store materials and make suggestions as to a properly fenced off and alarmed yard. Determine where a majority of the heavy equipment will be left at the end of the day so that it may be adequately patrolled and protected.

Keeping in mind that construction professionals have a multitude of signs, determine if there is any security-specific signage that your site will need. Most security companies have signage that identifies the company and has a telephone number on them. However, you may think of some site-specific signs that could be made to warn of the presence of your security officers. Although this may not deter a determined thief, signage can remove the argument that the crook did not realize that he or she was trespassing.

Concern yourself with the environment, noting whether it is in a densely populated area, a remote area, or somewhere in between. Don't be lulled into a false sense of security if your project is in the middle of nowhere. Thieves are highly mobile these days and are perpetually patrolling for something to steal. If a site is remote and the crooks think no one is watching, the site can suffer some very heavy losses from theft.

Determine where the high ground is for observation purposes and safety during flooding. Also identify the best places to fully observe a majority of the project, keeping in mind that your security officers will need to stop to rest and take their breaks from time to time.

If your security company or the contractor will be using cameras, determine the best place for them at the outset of the project, keeping in mind that your cameras will have to be moved from time to time as the project progresses.

Ascertain where the nearest law enforcement agency is and obtain the address and telephone number. Calculate the approximate arrival time of the police for consideration when planning how long security officers might have to detain suspects. Make a visit to that agency and speak with the community relations officer or watch sergeant so that you can begin to build a relationship with law enforcement. Ask about crime trends in the area and ask if they have had construction sites in the area recently. The answer to these questions will help you plan your operation by being aware of criminal activity.

Make a similar study into response times for the fire department and paramedics. This information is important not only for security officers but also for everyone who will be working at the site. Make sure to find out if you are in a critical fire danger area, which will help you ascertain of you have the need for firefighting equipment on site.

If duties permit, make a visit to the site at night. I can't think of a single place on Earth that doesn't look different at night. Things that are clearly visible during the day may be completely invisible at night. If there are streetlights, businesses, or residences nearby, determine if those light sources are of any use at all to your security operation. It's okay to use existing light sources to help, but relying on them can lead to failure if they are not bright enough.

Because you will already know where your equipment and materials will be placed (for the most part), determine where you will need additional lighting. This is critical because light is vital for your security officers and is also the enemy of thieves. Light will add cost to the construction project, and it will be incumbent upon you to convince the site manager that there will be a return on that investment if even one major theft is prevented.

SITE SURVEY AND VULNERABILITY REPORT

Prepare a report detailing your findings and recommendations for the client, as well as the security management personnel who will be overseeing the project. Start with an opening paragraph or two detailing the scope of your survey. Then pick a logical way to wind your way through the project so that the reader can follow your progress easily on a map or on the construction plans.

In today's digital world, it is easy to add your photographs to your report. This is not for fluff or to make your report longer but rather to illustrate the security needs that you will be discussing in this document. The old adage that "a picture is worth a thousand words" is true because with the right images, you can convey the importance of your security points. Consider the use of maps in your report as well. A properly marked map can clarify exactly which area you are discussing in your report.

Use your research of the area and your discussion with law enforcement and fire professionals as support when you discuss the risk and vulnerabilities associated with the site and your suggestions for risk mitigation.

Finally, list each recommendation clearly and succinctly, one by one. These should be easy to understand and logically justified. Make sure to state that as the site progresses, the security plan should be flexible and fluid, adaptable as the site changes.

This report should be accompanied by a face-to-face presentation with both construction and security management personnel. You may not get buy-in on all of your suggestions (usually because of cost), but you are more likely to have success on a majority of them if you are there in person to clarify the points made in your report.

THE INDISPENSABLE AFTER-ACTION REPORT

Again, the computer is valuable in today's world because you can keep your reports on file without the need for an excess of paper. You want to keep every report and document on file so that at the end of the project, you can ascertain the effectiveness of your security plan.

If you work for the security company or maintain contact with them through the course of the construction project, you should make arrangements to be advised of any security-related incidents that occur during the project. This will help you to amend the plan if needed and will help you and the client evaluate its effectiveness.

At the end of the project, you should prepare an after-action report, if only for your own edification. This report should include a synopsis of your original site assessment and vulnerabilities report and should highlight any significant developments during the project. You should include any changes to your security plan that were made out of necessity. You should detail the successes, the failures, the unexpected events, and the "near misses." This report will help you sharpen your skills in risk assessment and problem solving.

As an added bonus, these reports can keep you from having to "reinvent the wheel" every time you do a site assessment and security plan by refreshing your memory as to the plans that were successful.

Your construction clients should have a copy of this report. Some will look at it, and some won't. Don't get your feelings hurt if they don't read it. Keep in mind that their main preoccupation is building their projects. Many look at security as our job and will be satisfied if nothing major occurs during the project.

Your security clients (either your employer or your consultee) should have a copy of the after-action report. These people should most definitely be concerned with the contents of your report. They can use it to assess if any changes that they made to your plan had any impact on the job. They can also use it to help them plan future jobs by determining what worked and what didn't.

Last, you should always keep a copy of all of your reports for the reasons stated in this chapter. They represent a record of all of your hard analytical work.

CHAPTER POINTS BY DISCIPLINE

1. **Security professionals:** To the security professionals who are responsible for securing large outdoor capital projects, the site survey and vulnerability assessment are among the most important planning tools for your operation. If you operate without a plan, you are just "shooting in the dark." You might get lucky, and nothing will go wrong at your job. If so, you rolled the dice and won. Keep in mind that one of the goals of this book is to raise the standard in construction security.

    Even if you <u>have</u> a small company with only a few security officers, you can still conduct business in a professional manner. A good plan and its execution will lead to success, and a natural outcome of that success will be more business.

    Regardless whether your company is large or small, having a plan, evaluating that plan, and learning lessons from its execution is a formula for success. Having no plan is a plan for failure. Failure is not an option.

2. **Construction professionals:** You would not undertake any construction project without first planning in detail every step of the way. You write a proposal, and upon winning the contract, you buy materials, hire the right people, and get your equipment ready to go on the first day of operation. Considering the potential for loss of equipment and materials (read that: money and time) because of theft or other security-related incidents, why would you not plan in advance in the same detail for the security of your project?

    The point of this chapter is to encourage the mindset of construction professionals to accept the concept that a proper security plan based on a sound site assessment and vulnerability report is worth the investment. Insist on meeting with your security professionals before starting even if they don't ask for a meeting. After all, you ultimately will decide what parts of the plan to accept. This meeting is also your chance to proffer your security experiences during jobs and tell the security professionals what has worked and what has not. If you are hiring a small security company, your experience in such matters may eclipse theirs.

    Insist on a site assessment and vulnerabilities assessment and, after the project is over, insist on an after-action report form your security providers. Share the results of the report with your crew so that they, too, will develop a strong sense for security.

3. **Law enforcement professionals:** If you are a detective or a patrol officer or if you are assigned to community relations, it is incumbent upon you to reach out to the construction companies and security companies in your jurisdiction and offer your assistance in the planning segment of the job. You have the unique

information about criminal activity in your area and can share the trends with the security planners.

If you are in the meetings with the security professionals who do the site assessment, your information could change the whole tone of the security plan. Their plan will be based on the best information that they can get about crime and traffic concerns in the area. Your input will help make their plan more accurate and thus more valuable. Ask for a copy of the security plan and follow the project as it moves along. The relationships that you develop in this process will help you achieve your goal of reducing crime in your area.

4. **Insurance professionals:** As in most cases, insurance professionals have the ability to drive activity in industry through the control of rates. Site surveys and vulnerability assessments make sense when used to develop a security plan.

   If you are in a position to compare the survey with the actual plan that is implemented, you can evaluate if the security for that site is adequate to protect the assets that you are covering.

   If your companies have CPPs on staff, they can assist you in evaluating the security plans for your sites. If you avail yourselves of the after-action reports, you can use that information to help advise your clients on effective approaches to site security.

5. **Legal professionals:** Regardless whether criminal or civil litigation is initiated because of an incident at a construction site, it is potentially valuable to ascertain if security at the site was adequate to deter criminal activity or other tortious incidents.

   Because none of these surveys, assessments, and plans are required by law and are entirely voluntary, they only go to reason and prudence in providing for security at a construction site. However, if you are in an advisory position at a corporation or regulatory agency, you might be able to effect change by requiring these surveys and assessment before approving a contract or granting permits.

Recovering After the Loss

6

CHAPTER OUTLINE

Given that losses will occur, this chapter teaches contractors and vendors how to plan in advance to get the job back on track. It teaches them that having an action plan can save them from excessive down time after a loss. It shows them how to prearrange for vital equipment, personnel, and materiel in emergencies and how to form mutual aid agreements with other companies to help one another recover after a significant loss.

PLANNING FOR THE LOSS

"The best laid plans of mice and men oft go awry."

—Robert Burns, Scottish poet, 1785

Robert Burns wrote this line in his poem, "To a Mouse," in 1785 after he had accidentally destroyed a mouse's nest while plowing a field with his brother. He had intended to plow the field, and the mouse had intended to shelter its family for the winter. Neither situation turned out as planned, although I suspect the plan went better for Robert Burns than for the mouse.

The point is that no matter how carefully you plan, the likelihood that something will go wrong is inevitable. The likelihood that something will go drastically wrong is impossible to predict, but rest assured, it's always looming out there, and you need to be prepared for it to minimize its impact. A catastrophe will slow you down and break your stride to be sure, but it doesn't have to destroy your project or put you out of business if you plan for the loss. There are rules that will keep you prepared for disaster and help you recover from your loss.

1. The first, most cardinal rule in recovering after a loss is **KEEP A MONETARY RESERVE**. Just to make things clear, **a credit line is not a reserve**. Credit lines can dry up without warning, especially during a disaster that impacts lots of people. Only you can determine how much should be in your reserve, but a general rule of thumb is to accumulate enough money to continue your operation for 3 or 4 months. It is not necessary or practical to set that money aside all at once unless you already have that money. It is perfectly okay to assign a percentage of your company's income (say 15%) to an interest-bearing account and **leave it there** until an emergency occurs. This reserve is not "buy-a-boat" money or "take-a-trip" money but rather transfusion money for your business. Money is the life blood of your business, and a mishap can cause your business to need a transfusion of funds.

2. The second cardinal rule is to **ALWAYS HAVE INSURANCE**. Have enough insurance with as high a total coverage as you can afford. Operating without insurance or with too little insurance is another form of "rolling the dice." In most cases, you won't be able to get the contract without insurance, but some people roll the amount of coverage back or drop insurance altogether after winning the contract. This is a formula for certain failure if something goes wrong. Those few pennies that you might save by rolling back insurance are not worth the losses you will suffer if something goes wrong.

3. The third cardinal rule is to **ALWAYS HAVE A GOOD ATTORNEY**. By good attorney, I mean a person who has actually studied the law and graduated from law school. If you don't already know this attorney and trust him or her, find one whom your friends recommend and check his or her references. If the attorney is legitimate, he or she won't mind your doing a background investigation. Check to see if the person's clients are satisfied with his or her services and fees.

Attorneys who actually know the law and practice it honorably can save you a ton of money when you need them. Pay them their fees gladly, without hesitation or delay. They are worth the money.

If you don't know anyone who knows an attorney, contact your local bar association. They can generally provide you a list of trustworthy lawyers who will serve you well. Research two or three of the best, interview them in person, and then pick one. Most attorneys don't charge for the first interview, especially if they know that there is a long-term business relationship ahead.

If you are a small business and think you can't afford an attorney, consider that one unprotected serious incident can put you right out of business, and instead of growing your business, you will have to go out and work for someone else again.

ACTION PLAN

During emergencies, most people I know who "shoot from the hip" begin to act seemingly without thinking. There is nothing wrong with that if you know how to shoot. Those who are successful at this emergency management generally have already thought about it before the emergency. They seem to automatically know what to do. Why is that? That is because they **have a plan**.

So, where does one start with a plan? First, begin with the routine things that can go wrong during a work day (e.g., equipment failure, absence of key personnel, scheduling errors, failure of vendors to deliver needed materials). Then make yourself a small checklist of what you will do if these things happen. After you have committed the small solutions to writing, you will remember them and may never have to look at your checklist again.

I have been a list maker all my life, and it used to drive my wife nuts when we were first married. I even made a checklist for our honeymoon trip. I remember exasperated sighs in the background as I meticulously checked off things as they went into the suitcases. I pretended not to hear the sighs because I knew that if we got to Europe and something was missing, it would potentially mar our trip. Luckily (or maybe by design), nothing was forgotten, and we had a great honeymoon.

If you look at a blank police report form or any other report form, there are blanks that must be filled in. Why? Because the information in these blanks is important, and the blanks form a checklist for the information required in a report.

Short lists are easy to remember, but what about more complex situations? Shouldn't you devote even more attention to the major things that can go wrong? I believe that planning ahead always pays off, and by making a checklist or using an outline, a notebook, or any other memory jogger, you can begin to train yourself to respond before the need to respond arises.

Very importantly, you want to plan to pick yourself up after a loss and resume operation as soon as possible. This concept will keep you in business and maintain your reputation as a reliable, resilient company.

ADVANCE ARRANGEMENTS

While planning for loss recovery, consider what will be critical and most important to restore things to normal. Will it be equipment, machinery, vehicles, materials, personnel, or a combination of some or all of these things? Whatever it is, you need to identify it and make advanced arrangements to get it when things go wrong. In your plan, include the contact information for those who can help you and put those contacts in your cell phone, notebook, or other device that you keep with you at all times. You don't want to have to scratch your head, trying to remember where that information is.

If you are in the business of building outdoor capital projects, you most likely already have some amount of your own equipment and machinery. You probably own the most critical pieces and possibly rent the rest. That means you probably already have a business relationship with the owner of the rental company. If that relationship is good, then talk with the owner and try to get yourself put at the top of the list when it comes to getting equipment and machinery during an emergency that affects all the construction sites in your area.

Take a look at the equipment and see if some of your equipment might be able to serve "double duty" until the proper piece of equipment can be rented or purchased. Look through your storage area where you keep extra equipment. You might find that you have purchased two or three of the same thing over the years.

Both construction and security companies use vehicles in their work. If these vehicles are stolen, damaged, or destroyed, you need them to keep operating. If you do not have sufficient vehicles in the fleet, then renting them is the only other viable alternative. Make sure to spend the extra pennies and get the rental insurance for the vehicles you rent.

Often at construction sites, materials that are not properly secured are stolen, usually at the most critical time. Of course, you have your suppliers that you use all the time, but if they can't serve your needs in an emergency, make sure to identify other suppliers to whom you can turn in an emergency. Even if you have to drive some distance to acquire this material, it will be worth it if you are at a critical point or are up against a deadline to earn a bonus.

Perhaps the most critical element that can help you extricate yourself from a loss event is personnel. Without people, the machinery, equipment, and vehicles will just sit there and do nothing. If you have a large enough company, you can bring in people from other jobsites to help until the catastrophe is resolved and things are back to normal. If not, you should develop a plan to get help from somewhere in a hurry. I will discuss this in more detail under the heading Additional Personnel.

MUTUAL AID AGREEMENTS

As I mentioned at the beginning of this book, I come from the government sector (law enforcement) and moved over to the private sector after I retired. In law enforcement as well as most fire departments in the United States and other countries, there exist mutual aid pacts between agencies to come to the assistance of another jurisdiction in the case of extreme emergency. This concept works well and allows agencies to work together for a common goal, although I must confess, I don't know how the bean counters figure out how everyone gets paid in the end.

During my CPP class, the same mutual aid concept was posited, but it was presented to be between private companies to resolve problems in emergencies. It boggles the mind to think that two or more profit-driven enterprises might actually work to develop agreements to assist one another in time of crisis. That would mean that companies that are potentially in competition with one another for contracts would come together, meet, strategize, and come to agreements to mitigate loss to one company or the other (or both).

This concept is not without precedent. Before the sheriff's department, I worked in the airline industry, and it was not uncommon in those days to exchange aircraft parts with another company to prevent a critical flight delay. It all seemed to work back then because every company reciprocated with each other and replaced the borrowed parts as soon as they became available. I'm sure there must have been an accounting aspect to this comradery, but I was in flight operations and left those details to the bean counters.

So, where does one begin to achieve corporate mutual aid agreements? The process begins with determining if you yourself can do this and if you are willing to reach out to others to make things happen. Without this mindset, there is no reason to even begin.

Every industry has organizations. By broaching the topic of mutual aid at organization meetings, you can begin to stir up interest for the concept. It is to everyone's benefit to help each other in times of need. It's important to start the conversation.

Most professional leaders have colleague who occupy similar positions of leadership and decision making in other companies. These business friends are excellent candidates with whom to also begin dialogue on mutual aid agreements.

When you identify a receptive person or agency, reach out and begin with a face-to-face meeting to discuss how this agreement would work. From there, you can design your own program that fits the needs of both companies. If successful, then invite others to join.

Such agreements are by their nature voluntary, and the real test of success will occur when the first emergency incident arises. However, none of this will happen if the first step is never made. Such arrangements could bring a vital piece of equipment or the right person to your site when you need it the most.

ADDITIONAL PERSONNEL

As previously mentioned, personnel are the key component to your company's success. Without the right people, all of the best equipment in the world won't do you any good. Therefore, you have to have a strategy to replace key people or hire additional personnel to get things back on track. Again, you must have a plan.

In most countries, there is usually some sort of government employment development agency that helps unemployed people find work, even if it's only temporary employment. There are generally always some out-of-work construction workers who can use some money, and these agencies can help pair them up with your company. Just make sure to vet them before you bring them into your employ to make sure you are not bringing unwanted trouble into your organization.

In the United States, Goodwill Industries also has a job outreach program. A contact with them may bring unexpected good results when personnel are hard to locate. I'm sure that in most countries in the world, you will find similar private organizations that seek to pair job seekers with companies that are in need of good personnel. Again, just make sure to vet them.

In today's digital world, you can post immediate job openings on a variety of websites and get almost instant responses. (Craigslist and ZipRecruiter come to mind, but there are many other sites.)

If your job is in a union-controlled area, then by all means, contact the union hall and ask them to send you qualified candidates to interview.

When you identify your candidates, have them and your human resources people meet you at the site and conduct your interviews there so as not to interrupt your recovery process.

Whatever system that you use to hire additional personnel has to work when you are recovering. If you ask your Uncle Joe to send you some friends, and it works, then good for you. If not, try some of these other options and get the work started.

RECOVERING REPUTATION

Your company's reputation is something that you can't buy. You have to earn it, and that reputation is built job by job, over time. A few things can cause a company's reputation to be destroyed immediately. Most of them involve ethics. Either your company operates in an ethical manner or it does not. If you are not operating ethically, there is no plan in the world that can restore your reputation.

However, sometimes events happen that damage your reputation that are beyond your control, and generally, you can recover from these events if you acknowledge the event, make things right with your client, and avoid repeating the error.

An example of this type of reputation damage occurred when a security company overpromised the British government for security coverage for the Olympics in that country. The company was only able to provide approximately 60% of the security officers that it had promised and was generally considered a disaster in the security field. Because the company had generally given excellent, complete service throughout its long history, it was able to survive the incident, but several people had to leave the company before its reputation began to recover.

Similarly, in construction, reputations can be damaged by not providing promised results even if it is through no fault of your own. (Natural disasters are not generally part of this equation.) Theft of materials or an important piece of heavy machinery can cause construction delays that could tarnish your reputation for timely completion of projects. If you recover quickly enough, you may not suffer any damage to your reputation at all.

If it looks like there may be delays, then contact your client and explain to them honestly and forthrightly what has occurred. Give them a recovery timeline and stick to that timeline. Everyone understands that bad things happen, and if you are open about an incident, your client will appreciate it.

After such a loss, of course, contact your insurance agency immediately. Then go out and acquire the replacement equipment that you had planned for before the incident and begin work anew.

Recovering after a loss, although inconvenient and costly, should not be difficult if you have planned in advance.

CHAPTER POINTS BY DISCIPLINE

1. **Security professionals:** The principles in this chapter apply to all businesses, not just the construction industry. The variables may be different, but the techniques should be the same.

    Keep in mind that, as a security company that guards outdoor construction projects, your role is to be a partner with your clients to help them at all times, even when the going is tough. Your client may be expert at building capital projects but may have been lucky enough to have never had a serious loss event.

    In that case, you as security managers should stand by your client and offer help if he or she is amenable to it. Recovery is a matter of common sense, but often in emergencies, common sense flies out the window. You can be with your client in an advisory role and be of great assistance. Remember, your duty to your client goes beyond just guarding their things. Helping them through a crisis and guarding their reputation is also part of your responsibilities.

2. **Construction professionals:** Hopefully, your losses will always be small. Losses will occur, to be sure, but if you plan in advance, they need not seriously interfere with your work flow. Take the time to make checklists in advance and keep them on your cell phone or other device so that you don't have to hunt for them when you need them. Chances are that if you use checklists, you won't forget important details in a loss event.

3. **Law enforcement professionals:** The chances are that during a major loss event at an outdoor construction site, law enforcement will be involved in some way. Of course, your primary duties are to prevent loss of life and apprehend criminals if a crime has occurred.

    However, if you have already partnered with the construction professionals at the beginning of this project and know them, you will more easily extract from them the information you need to perform your job because you will already have a relationship with them, and they will trust you.

    It is also important that you understand the enormous impact a serious loss event can be to a private company. Your professionalism as you attend to your duties during such an event can give the people who you serve in the construction business great confidence. As you already know, sometimes people just want to tell their story to a cop. If you understand the principles of this chapter, you can offer counsel to the construction professional when his or her mind might be clouded by the impact of a serious loss.

4. **Insurance professionals:** If you are an adjuster, you already know the importance of being there for your client during these critical emergencies, and hopefully you will be right there with him or her as soon as you become aware of a serious loss event. Just the awareness that someone "has their back" can help them focus on what they need to do to recover after a loss.

    It can be helpful to provide them with the checklists and information in this book, in addition to the helpful information that your company provides to its clients.

5. **Legal professionals:** As legal professionals, particularly as retained counsel for the construction clients, it is important that you keep on file your client's emergency plans. Of course, this is important because it goes to reason and prudence in your client's approach to business. This could be of important to protect your client during subsequent litigation.

   Furthermore, providing your clients with some of the checklists and other information in this book may help them to do the right thing when a loss event occurs, thus making your job easier.

Loss Mitigation Strategies and Tactics

7

CHAPTER OUTLINE

This chapter gives instruction to security and construction professionals on tactical plans and strategies to protect large areas with a minimum of personnel, vehicles, and equipment. It teaches that in most instances, construction professionals will spend the bare minimum to get by, and it teaches them how to present the case for better security to the client. Examples are used of a variety of outdoor venues, including maps and photos, to be used as exemplars for developing protection strategies and tactics. The concentration of construction equipment into protectable groups is discussed, keeping in mind that although this tactic is sound, theft from equipment is easier if no guards are present.

SECURITY CONTRACTOR VERSUS PROPRIETARY SECURITY

When planning to secure a construction site, decision makers must first decide whether to provide their own security or to hire outside security contractors. To be sure, for purposes of directing traffic around a closed road or other traffic impediment, it makes sense to use proprietary personnel who are otherwise unoccupied at the moment to direct and control traffic. That saves you money.

However, when planning the overall security plan for the project, consideration has to be given to using a professional security officer contractor in lieu of relying on site workers to be alert. Often construction firms rely on workers who might be camping overnight to provide nighttime security. After all, why pay someone else when you have people on site?

First, construction workers who have toiled all day on the project are likely fatigued and therefore are not as alert as needed to provide the necessary attention to detail. Second, most construction workers lack security training, and their use as security officers is a stopgap measure at best. The exception to this is if the construction company hires personnel whose sole duty it is to provide security for the site and those personnel have proper training in asset protection. If a company decides to use proprietary security, then it is imperative to hire a dedicated security supervisor (or manager, depending on the size of the project) so that the process of construction can proceed with a minimum of distraction.

Using a contracted security provider is generally preferable because security is what they do. Their service includes not only the security officers but also the support staff, supervision, equipment, and security vehicles designed specifically for patrol duties. You wouldn't hire a cardiologist to do a tooth extraction. The same concept applies to security. Their management should be trained in site security planning, and their personnel should have all of the required training and licensing to secure the site professionally and lawfully.

SELECTING YOUR SECURITY SERVICES CONTRACTOR

Selecting the right security services provider is of paramount importance to mitigate or prevent theft at the construction site. The size of the security services company is relatively unimportant as long as they provide professional service. Many local providers can provide you with exactly what you need. Often construction companies have long-term relationships with a local provider that meets their needs, and that is great. However, some construction companies feel the need to change security service frequently, which is most likely a result of not identifying the right provider at the outset of their projects. Also, many (but not all) large security companies provide excellent service as well, so you need to look at who will serve your needs well at a reasonable price.

So, what should you look for? The first thing you should look at is how the company's representatives approach you for the contract. I don't mean by this that they come at you like hungry wolves with their tongues hanging out, looking for as much money as they can get. Rather, I mean, what is their approach to serving you? Do they first present how their company operates and is managed? Do they talk about the equipment and personnel? What are their hiring and training programs like? Are they properly insured? What is their billing program like? Do they ask to look at your construction plans? Do they ask to walk the proposed site? Do they discuss the security equipment needed and ask about what you may already have in place? Do they just show up at your office with a clipboard and ask you what you want? I will answer these questions one by one.

1. **What is their approach to serving you?** As I said before, if you have a long-term provider, you work well together, and their efforts have mitigated loss at

your sites, perhaps this question is not particularly relevant to you. However, if you are seeking security services for the first time or are dissatisfied with your current provider, consider how the first contact with the prospective security provider goes and how confident you feel at the end of the presentation. Just as building stuff is more than just a job to you, you should feel a sense of enthusiasm about security services from the presenter.

Yes, you should understand that their presentation is a sales pitch, but it is no different really from your proposal for a job. You should feel as confident about their ability to provide good security services as you do in your own ability to build a great concrete project. Everyone has a different approach, and there is no one specific way to sell security services, but you should come away from that first contact with a feeling that the company can meet or exceed your company's security needs.

2. **Do they present how their company operates and is managed?** It is important to know how to expect that your security will be managed. If the company is a local or large provider, you should learn from their presentation where they are located and how you will be able to contact their management when there is a problem. You should learn if their hiring is locally based to ensure quick response to the site when needed. Keeping in mind that "you get what you *in*spect, not what you *ex*pect," they should give you a detailed explanation of their supervision program. They should also give you information as to their routine reporting procedures and how you will be notified during emergencies.

3. **Do they talk about their equipment and personnel?** Your potential provider should talk about such equipment as communications equipment, safety equipment, vehicles, tracking equipment, and inspection point systems. Tracking is particularly important because you want to know how they ensure that their officers are on duty and performing their tasks.

You need to know about their personnel, their numbers, and if you can always rely on a security officer being on post during the required hours. You should also ascertain the availability of additional personnel if they are required during special situations or emergencies.

4. **What are their hiring and training programs like?** You might be surprised to know that in many jurisdictions throughout the world, there are absolutely no requirements that security officers be trained or licensed. Even in the United States, several states have no licensing or training programs for security officers. That being the case in those places, it is doubly important that you know about hiring and training. Why? Because you don't want to end up with a site full of unscreened ex-felons watching your valuable machinery and materials.

The first thing you need to know is how the service provider recruits and screens its security officers. In places where no professional certification is required, some security companies have been known to go out to the highways and byways and literally pick up anyone who looks like they might be willing to work. Needless to say, this is a formula for disaster.

Recruiting is the first thing to look at. Do they use the "highway and byway" method, or do they recruit using advertising, through government agencies, job fairs, or security officer schools? What training and experience do they require of their candidates? Do they hire military veterans? Regardless of the country in the world, veterans seem to have a high application to duties and display good discipline.

Then you need to ascertain what their hire-screening program looks like. Do they check for a criminal record and credit history? Do they require drug screening? Do they actually check references, or do they just take the applicant's word for it that he or she is a good worker?

Last, you should check into their training program, if one exists. Some jurisdictions require a minimum of initial training hours, combined with a standardized set of required yearly training to maintain the security officer's license or rating. (California is one of these jurisdictions. Information on California's training criteria can be viewed at www.bsis.ca.gov.) Some jurisdictions require absolutely no training whatsoever. Therefore, if the prospective security contractor does not provide its own training, then you might be getting a human scarecrow with a costume and a plastic badge to guard your very expensive equipment and material.

5. **Are they properly insured?** It is totally appropriate and judicious to ask your prospective security contractor to provide you with proof of insurance. Any security company that operates without insurance should be completely avoided without reserve. When you consider vicarious liability, you can rest assured that if their security officer commits an actionable tort (civil wrong) against a third party that ends up in a lawsuit, the trial lawyers will reach right through the security company's empty pockets and stick their grubby hands into your company's deep pockets to extract money.

   A quick consultation with your insurance agent or corporate attorney will convince you of the need to require proof of insurance before you allow a company's security officer to set foot on your site.

6. **What is their billing program like?** Most security officer companies bill weekly with a net 10-day payment due date. Some larger companies can go net 30 day, but few go beyond that figure. The reason is that the security business is labor intensive and requires that money quickly to pay for their program. The smaller the company, the more frenetic bill collecting efforts will be. Otherwise, the owner has to pay salary out of his or her own pocket, and if he or she doesn't have the money or can't borrow it, you might find your site bereft of security officers.

7. **Do they ask to look at your at your construction plans?** This is a key indicator to see if the prospective company is professional or amateur. They should be very curious to see what it is that you are building so they can provide you with a good site assessment before the start of service. That is not to say that they will be experts at blueprint interpretation. That is your job. However, they should be able to look at your plans and schedule and make recommendations for optimum security coverage for your project.

8. **Do they ask to walk the proposed site?** This is another key indicator of the prospective provider's professionalism. After all, how can they give you an accurate security plan and price proposal if they have not even physically looked at the property? If they ask to walk the site, you should either send a company representative who is familiar with the project and who can answer questions to walk it with them or go yourself. You should be able to tell by their questions if this is a pro-forma exercise or if they are asking the right questions.

9. **Do they discuss the security equipment needed and ask about what you may already have in place?** They should discuss with you what equipment, such as vehicles, cameras, communication equipment, and hard hats, that you require for the job. They should also inquire about your planned fencing, lighting, tracking systems, and camera equipment that you already have to assist with security.

10. **Do they just show up at your office with a clipboard and ask what you want?** Sadly, this is the approach of some security providers to their bids. What they are doing by this practice is throwing the responsibility of the security plan back on you. That way, when something goes wrong, and it will, they can point back at you and say, "Well, this is what you said you wanted." Although you may indeed know exactly what you want, they should demonstrate to you first that they know what they are doing.

THE IMPORTANCE OF PHOTOGRAPHS

Whether you are a security professional explaining the need for your security plan to a client, a construction professional explaining the site to your client, a law enforcement professional explaining your case to a deputy district attorney, an insurance professional evaluating a claim, or a legal professional preparing to go to trial, the importance of photographs cannot be understated.

In today's world of easily discernible cell phone digital photography, there is no excuse for not photographing every possible aspect of a construction site. The site should be completely covered with pictures during the initial site survey and at every significant step of the project. This includes photographs of the landscape, construction vehicles, fencing, security equipment and devices, and the project itself as it develops.

The initial photographs will help with Crime Prevention Through Environmental Design (CPTED) when preplanning your security operation. You should use your photographs combined with your site walk to suggest to your client ways to arrange the site so that an outsider can clearly see that there is a security system at work to protect the project.

During the course of this book, I have made and will make the case for a security program for every construction site from the very first day that the very first piece of equipment to be left there overnight arrives. Regardless whether the construction company provides its own security or hires outside security professionals to provide this service, the security plan photographs should be taken by the people who have been designated to provide coverage of the site throughout the project. They should be freely shared by the security provider with the client or vice versa and should be kept during and after the project is completed in case a criminal or civil case arises.

They should also be kept on file in case a security lapse occurs. They are most helpful in evaluating what may have gone wrong, which led to a theft, vandalism, or safety incident, and should be used in the after-action report that must be written after each major incident.

Of course, law enforcement, insurance, and legal professionals will want to take their own photographs for evaluation or evidentiary purposes after a serious event, but they should always ask the construction or security professionals to provide them with as many "before" photographs as possible. This will help those professionals to "see the scene" as it existed before the case.

THE IMPORTANCE OF CAMERA SYSTEMS

It is important to know that camera systems are *not* a deterrent to crime. Criminals know that a majority of camera systems are not being monitored in real time. I can't count the number of crooks that I've seen on video recordings of crime scenes. A prime example is the numerous videos of bank robbers caught on video. Very few bank robbers even attempt to hide their identities. So why then do banks spend a fortune on the best, most clear camera systems available? The answer is that banks can use these images to help law enforcement catch and thus stop these criminals one at a time. In rare cases, arrests even result in the recovery of some of the stolen funds.

So how does this apply to construction security? It applies in the same manner, especially when the cameras are connected to a high-quality video recording system. With good cameras, you can capture more than just images of thieves and vandals. Often, you can capture images of the vehicles that the crooks are using while they rip you off. If the camera is good enough, occasionally you can capture license plate numbers and turn them over to law enforcement along with the images of the criminals. If detectives press hard enough on the criminals, you might even recover some of the items that were stolen from your jobsite.

Cameras can also help solve mysteries as to things going on at your site. Recently, Melvin Staples, CPP (the president and one of the owners of the company for which I used to work) and I were making a client visit at the Security Paving Norwalk site. This was one of the rare occasions when Jason Mattivi was not out in the field but rather in the site office. As we walked in, we saw Jason and his site manager Hani Jamaleddine hunched over the computer screen studying a video of the yard that was taken before our security officer had come on duty. Someone had stolen a fire hydrant water meter from out of one of the water trucks. The video showed us a man dressed in construction clothes, complete with a hard hat and a yellow safety vest, walk up as bold as brass to the truck, find the key, unlock the truck, take the meter, and walk off site. The camera was not set at the proper angle to see the suspect's vehicle, but the yard was across the street from a Walmart store, so Melvin and I went there to see if their store's outside had captured the image of the thief, or better yet, a license plate number of the vehicle that thief had used while perpetrating this crime. Even with the help of Walmart security personnel, we were unable to ascertain any further information on the thief.

Upon returning to the site and reexamining the video, Hani remarked that the man in the video resembled one of the site employees. Looking at the video, the man appeared to be wearing a Security Paving hardhat and safety vest. Shortly after that, it was discover that the "perpetrator" was indeed a Security Paving employee who had merely removed the meter from the water truck to place it in the fire hydrant across the street to fill another water truck. Melvin and I had even walked past the fire hydrant with the meter in it as we walked back and forth between the site and the Walmart and hadn't seen the meter, which was clearly attached to the hydrant. ...Mystery solved. Without that video, we would have thought the meter had been stolen, and Security Paving would have had to pay another $1400 to the Department of Water and Power to purchase a new meter.

One of the necessary additions to make cameras worthwhile at night is good lighting. Without proper lighting, you will have either poor video images or no images at all.

THE CRITICAL IMPORTANCE OF LIGHTING

As time progresses, lighting just gets better and better. Metal halide and now LED lighting have brought the brightness and color rendition that is needed to illuminate a variety of security applications at night.

As I said before, light is the enemy of thieves. Why do we leave a porch light on at night? So people will know that someone is home and awake. The concept means the same thing at your construction site at night. Great lighting is also important to help your security team see the machinery and material that they are protecting. It also makes what your camera system sees visible when the images are replayed.

Lighting at construction sites is moderately costly because it generally requires that you keep at least one generator running all night long. Generators require fuel, which is an expense. I hate to sound like a broken record (not really), but when you weigh the cost of one major theft against the cost of the fuel to power a generator, it makes good business sense.

There is also a safety element for the security officers who are site during the night. Not only does it allow them to see the equipment and material, but it also helps responding law enforcement to find the site in the dark if it is in a remote area.

Even if the cost of lighting is not built into the construction contract, being able to see your site at night is a "must" and should be a top priority in your security program.

WALKING THE SITE

Construction professionals wouldn't even think of starting a construction project without walking it first to get a clear picture in their mind's eye of how their plans match the landscape. The same goes for security professionals. This is a separate type of site walk from the type of walk construction personnel do, but the principles are the same.

First, you should begin with a conference with the site manager and ask him or her to show you the project plans and explain how things will progress during the job and what the planned dates are for completion for all phases of construction. This will give you a clear picture of the time frame for which you will need your security officers and will help you determine how many of them you should hire. If the site manager is amenable, schedule the site walk with him or her. They can point out things that may not be self-evident if you walk the site alone.

When you get to the site, again, *take pictures*. Take pictures of every feature of the site imaginable and use them to refer to when you are writing your security plan or proposal. Make drawings of things that you can't adequately capture in a photograph. Make note of where the site yard will be and ascertain how the construction company plans to fence in, light, and secure equipment and materials. Determine where it is likely that material and machinery be left overnight and think of how best to light that which is left exposed.

Identify the best observation spots for your security officers to station themselves, keeping in mind that these spots will most definitely change as the project progresses. I guarantee that the best spot in the world will become useless if a large concrete structure is erected, and you should plan in advance to identify future observation spots. That is one reason that it is important to look at the site plans.

Identify the high ground where your security officers and construction personnel can retreat to in case of flooding. Also look at the soil and ask the construction professional what its consistency will be like during wet weather. It is possible to look at decomposing granite, for example, when the weather is dry and not realize that when wet, it becomes as slippery as an eel. With this information, you can determine what type of vehicle will be best for the project and what kind of tires it may require. If you are requiring your security officers to use their own vehicles, you can devise criteria for when to patrol and when to stay put during inclement weather.

The site walk is an excellent time to test your communication equipment before the time it is actually needed. If the site is in a remote area, there is a least a chance that cellular service may not be available or only available in certain spots. Then you have to think about radio communication if you are in a line-of-site position to send and receive radio traffic. Communication is vital at any security guard site but is most critical when you leave a human being in the middle of nowhere in the middle of the night at a place where people are likely to conduct thefts or vandalism. Even if every piece of equipment and every pile of material remains untouched, we have lost if our security officers are harmed or killed because they could not call for help.

The site walk should also be conducted in part at low or no light. (Yes, that means you will have to stay out after dark.) As previously mentioned, things look different at night. Most likely your security officers will be there during the night, and you need to know what they will be facing. Also, any cop in the world can tell you that a whole host of different creatures comes out at night. It is the ideal time for cockroaches, both insect and human, to emerge and ply their trade. This will reinforce the need to supply your officers with the best lighting equipment available, including powerful flashlights, spotlights, and generator-lighted staging areas to deter the cockroaches.

Also, as part of the site walk, you will have identified in advance the nearest law enforcement and fire stations. You should drive to each station from the site. Drive at normal speeds but time the drive. That will give you an approximate idea of how long it will take for law enforcement or the fire department to get to your security officer in an emergency.

As soon as duties permit after all of this, you should prepare your vulnerability and risk assessment for the client, as well as the security plan.

PROTECTABLE GROUPING

Okay, so "protectable grouping" is not a very technical sounding concept name. However, I coined it because I thing it best describes how equipment and materials should be left at the end of the workday. As I mentioned earlier in the book, many construction companies leave their machinery and valuable materials where they lay at the end of the shift and walk away with every expectation that their stuff will be there the next working day. This is a concept that I would like to change in the industry.

I propose that construction sites be divided into zones that can be changed as the project progresses and that at the end of the day, designated movers drive the equipment to these staging areas and move the portable materials to the same place. Then portable lighting (lots of it) should be placed at each staging area. Yes, this will cost money, but it will also make each staging area a "protectable grouping" that can be secured with proper security procedures and the right security officers.

Keeping in mind that light is the enemy of thieves, this will be worth the expense and effort even if a special employee has to be appointed to make this happen. The theft of one huge piece of machinery or of a pile of steel or copper would more than offset the cost of such a plan.

Additionally, fencing off these protectable groupings is a safety decision that the construction professional will have to make, but I recommend it because fencing creates a deterring and delaying barrier for thieves. If vehicles are inside a fence, it becomes impractical to siphon gasoline and carry it away without having to cut the fence. Machines that are left in the open are highly susceptible not only to gasoline theft but also to battery theft, copper cable theft, and theft of the actual machines themselves. Lighted fencing staging areas delay thieves, increase their chances of being caught, and create a diminishing return on the effort to steal. Basically, if the site appears too risky, thieves will most likely move on to an easier target.

MULTIPLE OFFICER SITES

While I am in the process of suggesting additional costly security measures, I am going to posit that at some construction sites, the deployment of more than one security officer during a shift is wise. There, I've said it. So shoot me (Figure 7.1).

Again, I am going to hearken back to the bridge at Hesperia, California, discussed in Chapter 4. To refresh your memory, the bridge was built to connect the east side of Hesperia with the west side and give residents an additional freeway off-ramp to get home quicker. The bridge spanned 10 lanes (5 each way) and a huge median in addition to large shoulders (verges) on each side. The distance between sides at the beginning of the project was the better part of a mile. On a clear day, as the ramps were being graded, you could barely see the other side. At night, seeing the other side was impossible and was only slightly improved with lighting.

It took between 10 and 15 minutes to get from one side to the other by using freeway overpasses more than 1 mile each north and south of the post. Additionally, there was a site office on the south west side that was more than 1 mile from the project.

FIGURE 7.1

How far can you see?

Keep in mind that the entire area had a predictable set of nighttime denizens who were constantly prowling construction sites for something to steal and that there was equipment and materiel on both sides of the project. Also remember that in the offsite yard that was southwest of the project, we were asked to protect the entire project from the east side of the freeway without even using vehicles to patrol.

That came to a screeching halt one night when every battery was taken from the construction machines and their fuel had been drained on the west side while our officer on the east side was dutifully posted but unable to see. We were instantly allowed after that to begin patrolling both sides of the freeway, which worked until the crooks started watching our security officers and timing them as they made their random patrols. The thieves soon figured out that even if the security officers made quick post checks on either side of the freeway, there would still be an half hour in which they could conduct their thefts with impunity.

Shortly after that, the offsite storage yard (which we had not yet been assigned to patrol) was burglarized, and a very expensive atomic soil compaction measuring device was stolen from a shed, among many other valuable pieces of equipment (Figure 7.2).

FIGURE 7.2

The yard at Hesperia.

It became clear from that point that we needed one security officer for each side of the freeway. Yet despite all of these thefts, no additional officer was ever added to the post. Why? Because there was not money in the project's budget for a secondary security officer.

Although it was demonstrable that one officer on each side of the freeway, coupled with proper lighting and protectable grouping most likely would have mitigated some, if not all of the theft, there was simply no money in the budget for an additional officer and no enthusiasm for grouping and lighting the equipment where it could be seen.

One of the purposes of this book is to change that thinking. The cost of security should be a figure asked for in the Request for Proposal by the client, and should be built into the proposal by the construction company bid writer.

CHAPTER POINTS BY DISCIPLINE

1. **Security professionals:** It is incumbent on you to take the lead in security measures when it comes to bidding the job and performing your duties after you get the job. Keep in mind that your client's needs go beyond the sales meeting and subsequent proposal. If you haven't already done so, begin to think beyond just merely sticking a security officer, a car, and some equipment on the post and expecting the officer to actually deter crime. Begin to develop strategies that will show potential thieves that your site is actually protected. Even if your client doesn't buy into all of your plans at first, eventually he or she will as long as you can demonstrate the effectiveness of your security programs.

   Use the information in this chapter to more effectively assess your jobs before they begin, to demonstrate your company's professionalism, and to win the bid and grow your company. As you practice these concepts, your company's reputation as a reliable security provider will grow, and you will prosper.

   Join the American Society of Industrial Security (ASIS) International and participate in local chapter meetings and training sessions. Consider taking either the PSP (Physical Security Professional) or CPP (Certified Protection Professional) certification course. Learn the material and become certified. These certifications will go well beyond having a few extra letters after your name. They will provide you with some critical thinking strategies that will greatly assist you in the security profession.

2. **Construction professionals:** Begin to use these loss mitigation strategies as you bid your contracts. You should be able to demonstrate to your prospective clients how a good security program will help ensure a timely or early completion of the project. Begin to think of a security program as an integral part of every job that you have. This concept will add more considerations to what you are doing, so hire a professional to manage this part of your operation.

Choose at the beginning of each project whether you want to use your own security personnel, contract out, or a combination of both. Just make sure to provide your proprietary security people with the training that they will need. You might even consider sending them to a short security guard class and get them licensed for security with your local authorities.

If you choose to contract for your security and you don't already have a working relationship with a security officer service, get the word out to the security companies in your area about your upcoming project. I guarantee there will be a number of bidders.

Take the time to interview each bidder and gauge them on their ability to provide the service that you need. Go beyond the brochures and ask them about the security issues presented in this chapter. It will only take a few pointed questions for you to determine if they know what they are doing.

Become involved with the security program at the outset of your project. If the security manager suggests additional security measures, don't dismiss them out of hand. Consider how the suggestions may help you get to the end of your project without any major theft or vandalism incidents and go with them even if they cost a few extra pennies.

At least consider the "protectable grouping" concept presented in this chapter. This requires an additional manpower expenditure in addition to the fencing, lighting, cameras, and security personnel costs. Try to build these costs into your proposal. At the very worst, you may have to bargain these funds away, but there is at least a chance that if you explain your security program to the client, they will approve the extra money.

3. **Law enforcement professionals:** The concepts in this chapter should be of interest to you if you are in any way involved in policing the area around a construction site or if you are a detective who investigates industrial theft.

    If you know these principles and come across an unsecured construction site in your area, perhaps you can share some of these concepts with the site manager. Every criminal event that doesn't happen will make your job easier. Too bad there are no statistics available for crimes that could have occurred but didn't because a construction site was properly secured.

    I also suggest that you find and join your local ASIS International chapter. I guarantee that they would be happy to have a law enforcement officer in their chapter, and you might make some contacts in the security business for the inevitable day when you retire. You will learn that there is a great deal more to the security field than the security officers that you see posted throughout the world. Your expertise as a cop will provide valuable insight about the legal world to security professionals.

4. **Insurance professionals:** As insurance professionals, you are very involved in preventing or at least mitigating loss. If you are responsible for construction accounts, then you should know the principles in this chapter.

You are in a unique position to strongly suggest to your clients who operate without a security program ways that they can operate to help prevent losses due to crime. Knowing these principles will help you to partner with your clients to help them succeed.

5. **Legal professionals:** Whether you are a civil attorney, a criminal attorney, or a lawmaker, knowing the principles in this book will assist you in understanding the security principles involved at outdoor construction sites. Whether you litigate or legislate, knowing this information will assist to with your work.

If you are a counselor for a construction professional, you might want to assist your client in protecting their interests by properly securing their projects.

Gadgets

8

CHAPTER OUTLINE

This chapter highlights and showcases some of the equipment and devices that can and should be used to secure outdoor capital projects. It shows how they must be used in convergence with onsite security personnel. Site lighting, cameras, fencing, and gates are also discussed.

THE BRIDGE AT APPLE VALLEY

As I mentioned early in the book, I am not much of a gadget person. However, I admire the people who understand and design gadgets, and frequently I am reminded of their importance when securing outdoor capital projects.

Early one Monday morning, the first of December, I received a call from Mark Christie, the general manager of Security Paving, headquartered in Sunland, California. Because Mark seldom calls just to say hello, I knew something was amiss.

He was calling about a bridge project (Figure 8.1) that we were securing over the long Thanksgiving holiday weekend from Wednesday night around the clock until Monday morning.

This is a vehicle bridge that, when completed, will cross a dry riverbed, connecting two residential neighborhoods. It is about 1 mile long and is locked and gated at both ends with large chain link fences. That's the good news. The bad news is that

FIGURE 8.1

The bridge at Apple Valley.

other than the access roads, nothing else can be secured. This is horse country, and people have been riding in this riverbed for years. People also walk and run through the riverbed for exercise, and motor enthusiasts like to drive their all-terrain-vehicles through the dry wash and can approach from either end. Briefly stated, trying to secure this site is like trying to stop the water flowing out of a sieve. The best that can be done is to keep the gates locked and try to deter trespassers with verbal warnings. Luckily, our main job is to keep people from stealing the equipment and materials, which is relatively easy to do if your security officer is alert and has a telephone with which to call law enforcement.

It seems that during the course of this long weekend, person or persons unknown used a Caterpillar key to move a large scraper up to the south gate and then to move a large forklift up the dirt ramp toward the same gate. Unfortunately for the perpetrator(s), they managed to either drive the forklift over the asphalt berm while going up the hill or rolled backward over the berm. Either way, they high-centered the forklift on the berm and nearly turned it over. At that point, they abandoned their efforts and fled. Added to this mix was that one of the locks that secured the gates was a Caterpillar lock, and our security officers had a key to that lock on the key ring. The key that opens the Caterpillar lock can also be used to start any piece of

Caterpillar equipment in the world. (Remember that I lamented about universal locks earlier in this book?)

The problem for me was manifold. First, if we were manning the site all weekend long, how was it that none of our officers reported it to our command operations center, and why was it left for the client to discover on Monday morning?

Next, were we being set up for a large equipment theft, and if so, why didn't our security officer see or report any trucks or trailers in the area that were big enough to haul the machinery away? Did trespassers come up from the dry wash and drive the equipment around?

Where was our officer when this was going on, and why did he not report it? Was he complicit in the activity? Was he asleep or absent from the post? Was our security officer engaging in horseplay on shift, and did he drive the equipment around the site on a joyride?

What time did all of this occur? How could we prove it?

Mark Christie and the construction site managers were angry, and I don't blame them. They wanted answers, and I didn't have them—but I had to get them quickly.

Pacific Protection Services operates throughout California, and often it is difficult for one general manager to be in all parts of the state at once. That is why large security companies must have highly competent field supervisors, locally based and in close proximity to the jobsites. Lucky for me, one of my best and most conscientious field supervisors lived close (in California terms) to Apple Valley. I dispatched him to conduct an investigation.

Upon speaking to all four of the officers who were on duty over 5 days, three admitted to seeing the vehicles after the vehicles had been moved. The first, who was on duty in the middle of the night, said that he only saw the scraper by the fence but did not report it to his supervisor nor to dispatch. He did, however, make an entry in his daily activity report (DAR) and advised the officer who relieved him. The second officer admitted to seeing both vehicles, including the high-centered forklift, but failed to note it in his log or to notify anyone. The third officer, who came on at noon, saw both vehicles and noted their location and condition in his DAR. He then called dispatch, notified the dispatcher, and asked for his supervisor's phone number.

Upon calling his supervisor, he was unable to make personal contact, so he left a message. Unfortunately, because of bad reception in the area, the supervisor never received the message.

To make matters worse, the dispatcher either wasn't listening to the reporting officer or ignored his report; as a result, on Monday morning, it was the client reporting to me instead of the other way around. Things can happen on the construction site to be sure, but nothing makes us look like sleeping jackasses more than to not report something that should have been seen by the officer, and reported to the client in the moment.

Still, after all of the inquiry, all we knew was that three of our officers had seen the moved equipment, yet only one had properly reported it, and even that ball got dropped by dispatch. What to do?

In rides a gadget to the rescue. Mark Christie called and advised me that all of his equipment at the site was equipped with LoJack tracking devices. He was able to provide me with not only the times that the equipment had been moved but also with a graphic depiction of their movements about the construction site. The outcome was that in all probability, the first officer who noticed that the equipment had been moved had most likely been the one to move it. The other two possibilities were that he had been asleep or that he was not there at all. Regardless, he was discharged. The other officer who failed to report received a written reprimand, as did the supervisor who had not trained his people properly.

The whole purpose of this anecdote is to show the value of technology in unwrapping a mystery and solving a problem. Had this equipment had actually been stolen, the chances of its having been recovered were very good because of the tracking devices.

CONVERGENCE

In recent years, the concept of convergence has arisen in the security industry. This concept has arisen as professionals from two camps finally decided to join one another in the effort to produce the best security programs for our clients. I call the camp that believes in machines and electronic devices as the best security program approach the "gadget camp" and the camp that believes in human beings only as the best security program approach the "people camp." Some security professionals still believe in one concept and dismiss the other, and both sides are equally passionate about their beliefs.

However, cooler heads finally prevailed in this field, and the great thinkers decided to come together and see if using both technology and human beings might work. The late Col. Kuljeet Singh, CPP, always sang the praises of the new concept of "convergence" in which gadgets and human beings came together to form a unified front against the common enemy of crime.

This chapter covers the "gadget" side of things. I don't use the term *gadget* in an attempt to minimize or denigrate the value of technology. On the contrary, these machines and electronic devices have already proved themselves to be invaluable in the detection and prevention of crime and in the recovery of stolen property.

CAMERAS

The choice to use cameras in security systems for outdoor construction projects is up to the construction company or to the security provider. There is no hard and fast rule as to whether or not to use cameras. However, I highly recommend the use of a good security camera system as a forensic investigative tool.

As previously mentioned, security cameras are *not* a deterrent to crime. They are useful as a forensic tool, and if properly placed, they can be used to monitor a host of things beyond security. They can help you to monitor the progress of the day's work from a remote site and can show you who is working and who is goofing off.

You should determine the scope of what you will be viewing and how well you want to see it. Multiple cameras are available for day and low light use, and infrared cameras are available for nighttime use. There are cameras that are fixed, and there are controllable pan–tilt–zoom cameras that move side to side and up and down and zoom in on your viewing target.

At a construction site, unless it is extremely compact, you will most likely need multiple cameras and probably a combination of fixed and pan–tilt–zoom cameras. Wireless cameras, particularly the network IP cameras, are most practical for construction sites and are good for a number of reasons. First, there are no cables to run all over the place, which could be snagged and torn by construction equipment unless buried in the ground. They are also generally easier to install and are specifically designed to transmit images over the internet. Using a good program, you can view what's going on at the site via computer any time you like, wherever you are.

You should make sure your cameras are designed for outdoor use to keep dust, moisture, and bugs from destroying them or obscuring their ability to view the area. It is advisable to occasionally task an employee with ensuring that the lenses to all of the cameras are clean. This is particularly important after rain and after a dusty workday.

Infrared cameras are available to record what is going on in low or no light. They work off heat rather than light and are valuable to determine the number of perpetrators to a theft event and the time the event occurred. However, beyond that, their evidentiary value is low. If the images are used in conjunction with an arrest on site, they can be of some use to law enforcement and prosecutors. Otherwise, they just help you figure out what happened, when it happened, and how many crooks there were. You are better off with a high-definition camera combined with good color rendering lighting.

Before running to the hardware store and buying whatever they have on sale, it is advisable to contact an expert on cameras and camera systems and tell him or her what you need to view and why. Remember that these people, in addition to having an expertise in their field, are salespeople as well. Make sure you don't but 20 cameras when 10 will do.

Also be aware of **specialized cameras** and do your research either through your vendor, a trusted expert, or on your own if you have the expertise. Almost everyone who is involved in plumbing today is aware of sewer cameras and bore scopes. Among the new gadgets, there are also cameras that hang on copper pipes, are motion activated, and photograph pipe thieves in the middle of their criminal activities. For every security job that you can think of, there is probably a camera either designed for or adaptable to your needs.

LIGHTING

Lighting of outdoor construction sites is an essential part of any security program for your project. You can have the best cameras and fencing combined with highly trained security officers, and they won't do you much good at night without the proper lighting. Why? Because you need to see who is creeping around at night, and you can't do that without excellent lighting. That seems like common sense, but often construction professionals do not want to incur the fuel costs to run the generators that power the lighting. They believe that physical barriers are enough to keep thieves out. I assure you that physical barriers will not deter a professional, determined thief. They bring tools to defeat physical barriers, and although they will be delayed a bit, they will continue their criminal enterprise.

Every single area where there is machinery or material left out should be lit up brighter than the sun if possible. Every area of the site yard and every nook and cranny should be so well lit that there is hardly a shadow. So how do you do this?

Although metal halide, incandescent, and high-pressure sodium vapor lighting are available and give you varying degrees of luminance, the advent of LED lighting has given the world extremely brilliant luminance combined with a relative low requirement for electrical power. Because you are most likely running a generator to power your lighting system, you might ask what difference the power requirement makes. It means that you can cast a huge amount of light around every area of your project with minimal cost.

Not only will you save money on your lighting costs, but also in many areas of the world, the government will give your company rebates on the purchase of energy saving lighting devices. Furthermore, many of the outdoor LED security floodlights project usable light for a distance of 70 feet or more.

Another advantage to LED lighting is that many of them have a 100,000-hour life and, of course, require no light bulbs.

TRACKING DEVICES

Years ago when working for the Los Angeles Sheriff's Department, I became acquainted with LoJack, an excellent tracking device from an outstanding company. This was a new concept in that it represented a blend between the public and private sector in a most visible manner. LoJack clients would buy the satellite tracking devices for their vehicles and equipment, and law enforcement would install the tracking equipment into radio cars. The concept was simple: LoJack clients would notify law enforcement when their protected vehicles were stolen. Law enforcement would notify LoJack, which would in turn, turn on the devices' transmitter. Officers in LoJack-equipped cars would receive an alert when a stolen vehicle passed by and would call for another LoJack-equipped vehicle, and together, they would triangulate the position of the stolen vehicle. The crooks would get caught, and the vehicles would get returned to their rightful owners.

This system, I learned was also used to track railroad cars that had been abandoned on a siding in the middle of nowhere and forgotten. It was also used to track construction equipment, but I never had any contact with that kind of tracking.

Years went by, and I somehow ended up in the security industry. One morning I received a phone call from Gus Anaya at Reyes Construction, asking us to start service at a bridge construction site on Firestone Blvd. in Downey. It seems that someone had stolen two of their Bobcats (small multi-use tractors) from the construction site.

Even though they engaged us to secure their construction site, there was a happy ending to this story. Both Bobcats were equipped with LoJack, and they were recovered the same day, 2 miles down the concrete wash. The thieves had disconnected the buckets but had not yet taken them away, so nothing was lost.

This is not a commercial for LoJack, although LoJack has a direct connection to most law enforcement agencies in the United States. Today, countless companies sell tracking devices and the monitoring systems that go with them. The point is that your valuable equipment and vehicles should be equipped with a concealed tracking device to make sure you get your stuff back.

Of particular note, it is often overlooked that the most often vehicles stolen from construction sites in the United States are pickup trucks. Definitely, every pickup truck should be equipped with a tracking device. I would further recommend that you advise construction workers who drive their own trucks to invest the money in a good tracking system. A pickup truck does not have to be new to be stolen. Remind them that their trucks are as important to them as the tools they carry to work and are worth protecting.

Furthermore, with tracking devices, you can track your equipment's movements around the site. Remember the story of the equipment that was moved around the bridge construction site at Apple Valley. We were able to determine that the security officer had been joyriding in the equipment based on the hours in which the equipment had been moved.

Also, if you suspect that an employee is using a company vehicle for personal use, you can either prove or disprove it through an effective tracking system.

Another recent use of small GPS tracking systems has been to track copper theft. A well-placed GPS device in a roll of copper wire or in a stack of copper pipe can be used to assist police in locating your stolen material and arresting a criminal. Just as with cameras, with today's nanotechnology, if you can think of a use for a tracking device, there is most likely one that will adapt to your needs.

FENCING AND GATES AND OTHER PHYSICAL BARRIERS

Fencing in security of outdoor construction sites can serve a variety of purposes from warning of a dangerous area to defining the borders of the construction site to acting as a physical barrier to keep trespassers and thieves out. Regardless of their specific purpose, they have one thing in common. They create a defensible space which signals to everyone that an area has been cordoned off for a reason.

It is up to the construction manager to decide which fencing is most appropriate for the occasion. The choice has to be made whether you are screening the construction activity from the public eye to avoid distraction or whether the aim is to actually prevent thieves from accessing valuable equipment and materials.

There are valid arguments to be made for either type of fencing. The first consideration is what you hope to accomplish. Next, you should consider how long your project will last and if the protected area will be manned with either construction or security personnel during the project.

Of course, you should always keep in mind local ordinances regarding fencing. Some local municipalities desire to keep their cities beautified and consider construction projects to be an eyesore. Other municipalities may consider a construction project to be a distraction to motorists. This can be true. When a driver is busy playing "looky-loo" at the activity at your site, he or she can drive right into the back of another vehicle, or worse, a pedestrian or bicyclist. In these cases, you must use fencing that has a **privacy screen**. Some of this fencing is lightweight, such as orange plastic safety barrier fencing, which is designed to alert pedestrians to the presence of a construction site.

Keep in mind that whereas **privacy screening** may well deter some issues, it can exacerbate other ones. Remember that privacy is a thief's best friend. When shielded from public view, a thief is free to dismantle and haul away anything he or she likes. This is why I consider fencing to be a "gadget." With the use of the gadget of privacy fencing, you must have other security measures in place so that you have a convergence of devices (e.g., lighting and cameras) and personnel (construction workers or security officers) to make your security plan more effective.

Chain link panel fencing goes up quickly and can be easily reconfigured as the shape of the job changes. It is installed on stationary bases that accommodate two panels per base and require no drilling of holes in the ground. They are secured to one another general by brackets at the top and bottom of each side pole. You can buy or rent it with or without privacy screening. This type of fencing is difficult to find with barbed wire or razor wire adaptability, so if you need to doubly secure something, you are most likely better off with regular chain link fencing.

Chain link fencing comes in large rolls and is secured to steel poles that are secured into the ground. These types of fences can be open on top or may be topped with barbed wire or razor wire. If your project is going to be in one place for a long time and not likely to be reconfigured, this type of fencing is optimal. Your decision to top it with barbed or razor wire should be based on what you are protecting (e.g., copper wiring or piping) and the security program that you have devised. If valuable, easily stolen items are left unattended, then you will definitely want to make them difficult to scale by topping them with defensive wire.

Concrete barriers are available in a variety of designs and are most advisable for safety when a danger exists of workers being struck by vehicular traffic. K rails and Jersey wall barriers are designed to deflect cars and trucks and keep them from causing mayhem and destruction to the people and equipment on the other side. They can be purchased both new and used or can be rented. Either way, they are of utmost value when placed as a barrier against traffic.

The decision to buy or rent fencing and barriers depends on the size of your operation and the ability to store them when they are not in use. If you have to rent space to store unused fencing and concrete barriers, then you should ask one of your accounting bean counters to calculate whether it is wiser to rent these items or own them. The cost either way should be calculated into your proposal.

ALARMS AND MOTION DETECTORS

When I worked for Pacific Protection Services, we were contracted to provide security for Security Paving's crusher at a road construction project for Granite Construction Company. The project was on the inland side of the 101 freeway in Ventura County, just south of the Santa Barbara County line.

I met Guadalupe Rodriguez-Garcia, the project foreman for Granite Construction, and asked him what kind of security problems he had had during this job. He told me that numerous copper thieves had entered the site under a bridge from the beach side of the freeway. They had cut up an expensive copper cable, but for some reason, they never hauled it away. They had also stolen quite a bit of equipment and materials from the site yard.

I asked him what he had done to protect the assets in the yard, and he showed me. They had erected a barbed wire–topped fence around the project and had further installed motion detectors for the lights and motion-sensing alarms. They had also made liaison with the California Highway Patrol and the Ventura County Sheriff's Department and got commitments to provide extra patrols for the site. The combination of all of those things reduced the thefts significantly, and with the addition of our security officer to protect the crusher, the project remained safe until it was completed. It all started with motion detectors and alarms.

Outdoor alarm systems are available in many configurations from complex customized systems to turnkey premade out of the box systems. The amount of money one can spend on these systems is limitless, but that shouldn't mean that you should spend more than you need. This is one of the reasons that I advocate moving machinery and materials to protectable groupings every night. The more centralized your things are, the easier they are to protect at the end of the day.

When you select an outdoor system that includes motion detectors, make sure that you get sensors that are able to discern such things as changes in light, motion by animals, and wind. You also need to get systems that will stand up to rain and direct sunlight.

Your alarm system should be tied in with your remote camera system so that you can check your site from your computer or other electronic device from wherever you are.

COMMUNICATION DEVICES

As previously mentioned, communication between construction managers and security personnel is of utmost importance. It is also critical that security management and dispatch personnel are able to quickly and effectively communicate with their security officers in the field. Lastly, communication between construction management and security management is essential to maintain effective security and good business relationships.

In times past, everyone relied on land line telephones and in some cases line-of-site radios to communicate. Luckily, modern electronics has expanded the ability to communicate to a variety of highly effective devices. The advent of the cellular telephone and its innumerable innovative capabilities and applications has changed the way the world communicates.

At the very least, every construction site should have a cell phone so that security personnel can report anomalies and call for help. The more advanced the cell phone, the greater service the phone can be to the operation. Countless applications are available today for iPhones and Android phones that can turn them into a mobile office for the management of security. Officers can use the phones to generate daily activity reports and serious incident reports. Many programs allow security officers to record individual checkpoints that they pass during patrols and allow them to photograph anomalies in the moment and include them in their reports. In some cases, they allow the officer to send them to both the security company and construction manager, thus speeding up the notification process.

These devices combined with Internet-based communication via email are essential to instantaneous interaction between the client and the security service provider.

SECURITY VEHICLES

The choice of patrol vehicles, if required, at a construction site is critical. The security manager must evaluate the site terrain and assess if the security vehicle(s) will be able to negotiate the ground in all types of weather. Unless the terrain is flat and dry all the time, small automobiles are generally not suited for this type of patrol work. If automobiles are used, they should have sufficient ground clearance to negotiate rough terrain.

Generally, trucks, sport utility vehicles (SUVs), and Jeep-type vehicles are the most suitable for this type of work. Furthermore, the additional feature of four-wheel drive can save hours of extraction time if a vehicle becomes stuck.

Security vehicles do not have to be new, but they should be serviceable and in good repair. They should be regularly inspected and maintained to ensure optimum job performance.

The drivers of the security vehicle should be able to provide to the security company a clean driver history record from their local department of motor vehicles, and the history should be kept on file. Their driving during business hours should be routinely monitored by field supervisors. They should be aware that proper operation of a vehicle is as critical to their continued employment as is showing up to work on time and performing their duties in a professional manner.

Vehicle insurance is an absolute must. I wouldn't drive a vehicle 2 inches without insurance that at the very minimum covers liability and collision costs. Anyone who has spent time in the security industry has most likely seen a relatively new vehicle destroyed in a collision.

Lastly, vehicles should be clearly marked as security vehicles (with appropriate clearly visible lettering and markings) and should have proper lighting devices (yellow caution lights and spotlights) as allowed by local law. All lights should be in proper working order, and security officers should be instructed in the proper use of these devices. Routine inspection of lights should be made to insure optimum safety for the officers and proper visibility of the vehicle for the general public.

CHAPTER POINTS BY DISCIPLINE

1. **Security professionals:** As a security professional, it should be in your scope of knowledge to have an awareness of every security device available to secure outdoor construction projects even if you are not in the business of selling such devices. Part of your value as a security provider is the ability to advise your clients of the devices they might need to effectively create secure environments at their construction sites. Not every construction professional is necessarily a security expert.

   This is yet another reason for you to join the American Society of Industrial Security (ASIS) International if you are not already a member. There are several curricula of security knowledge such as the CPP (Certified Protection Professional) and the PSP (Physical Security Professional), among a host of other training classes and programs to enhance your expertise as a security professional. The knowledge that you gain in these programs will set you apart from other security professionals in the industry.

2. **Construction professionals:** Whether you have been securing your sites for years or just thinking about starting a security plan, it is incumbent upon you to identify the equipment and devices that will assist you in protecting your projects from theft and other disorder.

   If you would rather concentrate on building your project, then find the best security provider available in your area and interview them to determine what devices and equipment they recommend to use in convergence with designated security personnel. Then take the step of following their counsel and invest the money in these cost-saving devices.

   Begin thinking of the security equipment that you will need as you write your proposal. If you already own the equipment, then so much the better, but if you have to replace or upgrade things, it can eat into your bottom line if you have not already put it into the budget.

3. **Law enforcement professionals:** Regardless of the size of your agency or jurisdiction, if you are involved in any way with policing construction sites or investigating large thefts from these projects, it pays to familiarize yourself with the available equipment and devices that are used to secure these places.

   If you have made the requisite liaisons with the construction and security professionals in your jurisdiction, you will have the relationships necessary to evaluate their security programs from a law enforcement perspective and offer advice if needed for improvements to their security plan.

The ASIS courses that I recommend to security professionals (CPP or PSP) are also of value to you as a law enforcement professional. They will give you a better idea of the goals of security as opposed to those of law enforcement. You will find that as you involve yourself with construction site security, it will make your job as a peace officer easier and will assist to in catching crooks.

4. **Insurance professionals:** The devices and equipment noted in this chapter will help you to evaluate your client's security plan before the loss occurs. Even if you are only able to look at the plan on paper before the project begins, you should look at the systems and equipment that are planned for the job. If you are in a position to make recommendations to your construction clients in advance of the job, then do so. It may save your company an untold amount of money.

    Both the CPP program and the PSP program, which are sponsored and certified by ASIS International, will give you vast insight into what is involved in securing these construction sites. If your goal is to increase your company's bottom line by preventing or mitigating loss, then I recommend that you gain as much knowledge as possible about the security industry. The things you will learn at ASIS International will assist you in just about every aspect of physical loss prevention and mitigation regardless of the industry that you are insuring.

5. **Legal professionals:** Whether you are a civil or criminal attorney, a legislator, or a judge, it is important to know what equipment is available to secure outdoor construction sites. It is important to understand the reasonableness of prevention measures as taken by the construction companies and their surrogates, the security companies.

    This will also help you in a criminal case to evaluate what effort that accused criminals may have put into the perpetrating of their acts. These efforts go to prove motive or premeditation and help you to prepare a case for prosecution or defense.

    Whereas your caseload or other work may preclude you from studying for the CPP or PSP certification for ASIS International, I still recommend that you join the organization and attend your local chapter meetings and other events. As an attorney, you would be extremely valuable to your chapter to help assure that security concepts and methodologies are applied in a legal and ethical manner, including the use of devices and equipment.

Establishing Standards

9

This chapter presents proposed standards for the security of large outdoor capital projects. The suggested standards include security procedures, Request for Proposal (RFP) budgeting for security, legal standards, and other standards that apply to the outdoor construction industry.

All of the information in this book is of little use if the principles are not applied universally. It is my hope that the practitioners of security in large outdoor capital projects take the practices in this book and make them the standard for securing all projects.

HAVE A PLAN

We have all read the old saw that "failing to plan is planning to fail." Sometimes old saws prove to be true, and in this case, it certainly is. It should be standard in the industry that clients insist on a security plan to be included in the Request for Quote (RFQ) or RFP. If the client does not ask for this, the cost of security will have to come out of the profit margin of the construction company. It is the intent of this book to change the mindset throughout the industry to recognize that security is as important to timely and efficient project completion as is having the right personnel, equipment, and machinery.

The construction manager should devise his or her own security plan, allow the security manager to draft the plan if time does not allow, or ideally work together with the security manager to create a security operation structure.

The security manager should have a plan devised for each proposed construction site and should be prepared to present it logically to the construction site manager. Having a defined written plan separates professional security providers from amateurs and enhances the company's reputation.

Regardless of who develops the security plan, the standard plan format should include a description of the jobsite, including plans and maps, and should indicate the best places to place cameras and lighting. It must include security personnel, if any, and what their duties and responsibilities are. It must include protectable groupings of equipment and materials along with instructions as to how best protect them.

The plan should be accompanied by a security planning meeting, which includes all stakeholders. That meeting should frankly discuss all probable security challenges and the proposed solutions and strategies to mitigate or eliminate them.

PROTECTABLE GROUPING

A new standard of protectable grouping should be adopted throughout the industry. This means that all equipment and materials, whenever feasible, should be grouped together, fenced in, and properly lighted in as few groupings as possible. These groupings should be protected by onsite security officers and monitored by a recording camera system.

The equipment and materials should be spaced apart enough for easy detection of intruders, both by cameras and by routine patrol by security officers. Each grouping should have sufficient security inspection points that can be electronically documented to ensure that security officers are making their rounds.

EQUIPMENT AND MATERIALS MARSHALLING

This standard requires assignment of construction site personnel who are skilled at operating machinery to be responsible at the end of the day for marshalling all equipment and moveable materials to their respective protectable grouping compound. They can and should be assisted by the work crews who should move their equipment as close as possible toward the closest grouping compound at the end of the shift.

A decision should be made by the construction site manager whether to have equipment and materials moved back in the morning by the construction crew or to have a morning marshalling crew as well.

Record keeping is part of this process. Records should indicate which pieces of equipment were moved where and by whom. The record should indicate the times equipment and materials were moved.

This marshalling standard requires extra capital outlay for personnel and should be added into the proposal and quote.

DIVIDED SITES

This standard requires that whenever a site is divided into parts by a roadway, railroad tracks, a body of water, or any other configuration that requires lengthy travel to get to the opposite side that security officers, security equipment, and protectable grouping compounds be provided for both sides of the project. The sides should be linked to a central monitoring system. A reliable system of communication should be in place for the officers to contact their dispatch headquarters, as well law enforcement and each other.

Again, this standard requires more capital outlay than in times past, but when you weigh the potential loss from one major theft event or preventable incident, the return on investment in security will be worth it.

MACHINERY SECURITY

As I stated earlier in the book, many construction machines either have no key or have universal keys that fit every machine produced by their manufacturer. This necessitates key control (even if some thieves may be in possession of universal keys), ignition disablers, and physical devices to deter theft.

Also keep in mind that many copper cables and batteries, along with machine parts and tools, are stolen from these machines. This is why it is important to group machinery in well-lit, camera-surveilled protectable groupings so that after-hours activity can be closely monitored. This prevents having a theft occur on one side of the site while the security officer is patrolling the other side.

The standard should be to have a documented central key check-in system for all machinery. This system should also be responsible for accounting of all other keys for gate locks and for storage sheds. This system can be a collateral duty for a site supervisor or other trusted employee, but it should never be "self-serve." If key security is self-serve, then there might as well not be a system.

All machinery that has a universal key or has no key should be further disabled in some manner. This disabling can either be accomplished by removing a critical component or by use of a wheel locking system. The goal of this process is to delay the criminals who want to steal your machinery. Every minute that they are on your site increases the chance of detection and apprehension.

MATERIALS STORAGE

The standard for materials storage should be to remove all materials (except sand and gravel) from the open areas of the site until they are needed again in the morning or at the beginning of a new week.

Storage areas should be located within or immediately adjacent to the protectable grouping compound. They should be fenced in and secured. High-value items should be stored in a locked steel storage shed, and all items in the storage areas should be accounted for with a check-in/check-out system. Part of the theft experience is the loss of valuable construction materials that often go unnoticed because of a lack of accounting. Theft should never become an acceptable cost of doing business.

EMPLOYEE SCREENING

I wrote about background investigations in Chapter 4. It is important enough to cover in this chapter as well. I realize that with this type of book, some readers will read selected chapters, and I don't want this concept to be overlooked.

Recently, I attended an American Society of Industrial Security (ASIS) International lecture about employee screening by Pamela Graham, a retired FBI Supervisory Special Agent. The most compelling concept that I carried away from her presentation was, "Be careful who you let in your house."

This concept is especially important, both in the construction world and security. A large project could have literally millions of dollars' worth of equipment and materials on site. Do you know the people whom you are trusting to get the job done? What have you done to check the *bona fides* of those who carry your company's name?

The standard of screening starts with the employee application. Consider it to be the worksheet for your background investigation.

You should first look at it for completeness. Are there unanswered questions? I remember an applicant for a security dispatcher who had "forgotten" to check the "Have you ever been convicted of a crime?" box on the application. During the applicant interview, I learned that he had been convicted for embezzlement and had spent 2 years in prison. Had I not asked the question, this individual would have gotten further along in our background screening process.

Under the completeness heading, look for gaps in employment record. Was the applicant really out of work, or had he or she been discharged from another position for cause? Ask questions and seek explanations for employment lapses. Some lapses in employment are due to legitimate reasons and should not be a deterrent to hiring someone, providing they are otherwise qualified.

It seems elementary but check to see if your applicant has the requisite experience to perform the job for which he or she is being hired. Sometimes you meet a candidate whom you like from the start. You develop a bond and hire the person based on that relationship that you have established and forget the rest of the screening process. If the candidate is qualified, then great. But what if the candidate passes the charm test but can't do the job or, worse, is a criminal? Remember that one of the characteristics of con artists is that they can charm their way into just about any situation.

One of the most overlooked features of the application is the references section. Again, if you like the applicant, it is tempting to hire him or her on the spot without taking the necessary steps. If you are tempted to do this, tell the person that you like him or her but that you must wait from your personnel or human resources department to check his or her references. Most of the time, applicants will indicate people as references whom they think will speak highly of them. However, if the person is a decent but marginal employee, that fact will generally come out in the reference check whether or not the reference likes him or her personally. Remember that the reference may believe that his or her reputation is at stake and will most likely give you an honest appraisal of the applicant.

Next, check your applicant's eligibility to work for your country. In the United States, this is relatively easy, and it should be in most countries. You don't need the headaches associated with hiring someone who is not in your country legally (i.e., fines and potential criminal liability).

Depending on the law in your jurisdiction, pick the best lawful way to check your applicant's criminal history. Then check the results of that investigation against the application to see if he or she has been truthful. Depending on what job the person is being hired for, a criminal conviction for some crimes may not be a deterrent to hire a person if the conviction has nothing to do with the work he or she will be doing. (This does not apply to security officers, who should have a clean record.) Make sure that if an applicant has a criminal history, it is not for crimes involving theft or embezzlement. The best indicator of future behavior in that regard is past performance. Don't let a thief in your house.

Also look at crimes involving substance abuse, particularly sales of narcotics. Drug sales and theft go hand in hand.

Require your applicant to provide you with a copy of their driving record from your jurisdiction's department of motor vehicles, particularly if he or she might be operating a motor vehicle or motorized piece of equipment. (Most likely, your insurance company will require this anyway.)

Even if the employee won't be driving, you should require this of your applicants anyway. There are performance indicators on a person's driving record, particularly multiple convictions for driving under the influence of alcohol or narcotics and for reckless driving.

Repeated convictions for drunk driving are indicators of risky behavior and failure to learn lessons. This candidate would likely have attendance problems and could show up drunk to work.

Reckless driving indicates that the applicant has a wanton disregard for his or her own safety and the safety of others. This is a potentially deadly trait for a construction site. You don't need thrill seekers operating heavy machinery or being around hazardous situations. It does not make good business sense.

Again, it seems obvious that you verify that your applicants have the requisite licenses that are required by your jurisdiction to perform the tasks that they will be doing. This is a must. Depending on your local jurisdiction, operating without proper licensing can result in a costly fine or even lead to your site being closed down by government inspectors.

This concept also applies to the security officers who will be protecting your site. Strangely, some jurisdictions have no licensure requirements for security officers, but many do. For example, in California, if a security officer is working without his or her California Guard Card, it will result in a citation for the security officer and a hefty fine for the officer's employer. Aside from training requirements, a strong benefit of government licensing of security officers is that there is a built-in criminal screening process before these licenses are issued.

A check of an applicant's credit history is important if the laws of your local jurisdiction allow it. It is not being unfairly invasive to investigate whether your applicant has a record of paying obligations in a timely manner or whether he or she is "buried" in debt. (Generally someone is buried in debt if he or she is unable to meet even the minimal payments.) Bankruptcies, especially repeated ones, can also be an indicator of problems that the applicant has with meeting obligations. (Before I get too deeply into this, there can be legitimate reasons that an applicant has financial difficulties, for example, periods of unemployment, serious illnesses in a family, or other catastrophic events. That is why you want to ask applicants to tell you the story behind their financial difficulties.)

If an applicant is buried in debt, seek an explanation of what the borrowed money was spent on. If it was spent on food because there was no money or medical supplies because there was insufficient money or government assistance, funeral costs, or other reasonable expenses for which there was not enough money, then the explanation should suffice. For this type of applicant, if you hire him or her, you can point the person to programs such as Dave Ramsey's Financial Peace University or other debt reduction programs to help them break the debt cycle. If the person is salvageable and you help him or her in this manner, you will most likely gain a loyal employee who is working with peace of mind.

If an employee's explanation for excessive debt is that he or she is a profligate spender who will run off to Las Vegas on a whim or buy lavish gifts for him- or herself or for others to impress them, then pick another candidate. These types of employees will always be a problem because their financial behaviors indicate a self-centered attitude. These individuals will be the type to be late for or absent from work, with endless excuses as to why. Also, they are the type of employees who are likely to ask for loans from you that may or may not ever be repaid.

Repeated bankruptcies indicate that the applicant is not likely to follow through with commitments and will likely take the "easy way out" whenever it suits him or her. Remember again that the best indicator of future behavior is past performance.

Keep in mind that if you are seeking government contract, your government may require an additional background check on top of the screening that you did for the initial hiring phase. You can save yourself a lot of grief and potentially having to discharge someone you already hired by doing a thorough prehiring background check.

A strong screening program, regardless of what your company does, is one of the best ways to protect your organization from theft and other problems from within. Remember Pamela Graham's admonition, "Be careful who you let into your house." If you are not careful at the beginning, you will spend a lot of time and money cleaning up the mess later.

ACCESS CONTROL

Outdoor construction sites are generally out in the wide open spaces. Often, it is difficult to control who comes on and off the site by virtue of the fact that most construction sites are at least partially open to the public. The question then becomes, how do you keep unauthorized people out of your site? Even if the site is fenced in, this might not be as easy as it would seem.

This begins with knowing who has authority to be on the site. Schedules of workers should be distributed, and those who are to work on a given day must first sign in at a central location. Those who work for the company but who are not scheduled to work on a given day should check in at the same location and explain why they are on site. (If a major incident happens and nonscheduled personnel show up on their own, they should be approved by a manager or supervisor and allowed to work.)

Unscheduled personnel who frequently show up without a reason might be involved in something that requires their presence even when not on duty. This can signal involvement in a theft scheme or other embezzlement activity. Insider thieves often feel the need to be at work to cover up what they are doing. They may appear to be "workaholics" who just can't stay away and might even volunteer to work for free. Some employers will let this activity go unchallenged, but this is a mistake. Everyone needs personal time, and an employee who hangs around the jobsite and works without compensation is being compensated in some other way. Watch out!

So, how do you identify who is supposed to be at work at a glance from halfway across the site? A simple solution is to maintain control of some highly visible outer garment (i.e., helmet, safety vest, or company jacket) at the site. Each working employee must check out this item at the end of the day and check it back in at the end of the day.

Photo ID cards are advisable for large sites where multiple workers and outside contractors are on site who might not be known to managers and supervisors. These cards require vetting of contractors before allowing them on site. Outside contractors should have a different color ID badge than company employees.

A company culture of challenging strangers is a positive step to good access control. This doesn't mean a "Halt! Who goes there?" approach. This can be done in a friendly manner. Jim Grayson has a great approach. He advocates telling your employees to approach unknown personnel with a simple "Hi. You look a little lost. Can I help you?" The fact is that the person might indeed be a legitimate visitor, and as a result of your greeting, you won't have offended them. The visitor, whether authorized or not, can be directed to the site office. If the person leaves or does not comply, then you know that you have a trespasser.

Make sure you have signage at the entrance to the site that instructs visitors to the site office. Also, have someone who is specifically assigned to visitor and employee access control and keep a log of who comes onto the site.

CRIME STATISTICS FOR CONSTRUCTION SITES

This book seeks to establish statistical standards beyond those that involve solely the theft of machinery and metals. If you recall at the beginning of the book, I lamented that there were few crime statistics for theft and other incidents at constructions sites. I honestly believed when I undertook writing this book that I would find a treasure trove of statistical data about construction site crime that would help me to easily demonstrate the scope of this problem. I was shocked and dismayed to find that there appeared to be no centralized government statistical data base on these crimes anywhere in the world. To be sure, there are statistics on metal thefts but no indicators as to from where the thefts occur. Construction theft is such a universal problem that I thought every country in the world would be compiling statistics on these crimes, but sadly, that is not even remotely true. There are other crimes that occur at construction sites, such as assaults, vehicle collisions, and arson etcetera, that are also not captured in conjunction with construction sites.

The first statistical information that I found to do with construction theft was the excellent statistics kept by the LoJack Corporation. These statistics are specific to the recovery rates of equipment and vehicles that use LoJack's satellite tracking system. They were an honest evaluation of the effectiveness of their product. However, that was all I could find at first.

Then I read that in Great Britain, someone had formed a private and voluntary National Equipment Register (NER). The organization arose out of the need to be able to track stolen construction equipment. Even though some (not all) construction equipment have Vehicle Identification Numbers (VINs), they are not required by law in most jurisdictions to be registered with the government. I'm not opposed to that, as long as there is some mechanism to deal with registration. Enter the private sector, thank goodness.

I was pleased to learn that that concept spread to the Untied States. The NER here offers registration of construction and other heavy machinery into a central database. If members' equipment do not have VINs, this organization assists with creating registration numbers and helping owners to inscribe the numbers on their machinery. They also provide training to law enforcement specific to theft and recovery of heavy machinery.

One of the most admirable features of the NER is that it produces an annual report of statistical data for theft of heavy machinery. This report includes the type of machinery and where and when it is most likely to be stolen and recovered. These statistics apply to all machinery, however, and are not broken down by construction sites specifically. Having said that, hooray for the NER in the United Kingdom and the United States.

Without site-specific statistics, however, metal theft is just metal theft, and heavy machinery theft is likewise the same. By not grouping statistics by theft from construction sites, it is impossible to gauge the scope of the problem.

Therefore, I am positing that as a new international crime statistic standard, law enforcement statisticians begin keeping separate discernible statistics for all crimes and traffic collisions occurring at construction sites.

These statistics should have a general heading of "Construction Site Crime and Vehicle Collisions" and the be further broken down into categories such as outdoor construction sites, high-rise sites, commercial sites, housing tracts, defense sites, miscellaneous sites, and any other type of site that law enforcement professionals can think of.

The statistics should be then broken down into theft of heavy construction machinery, vehicle theft, equipment theft, metal theft and theft of materials from construction sites, assaults, arson, and vehicle collisions (fatal and nonfatal).

Furthermore, follow-up statistics should be kept as to recovery rates and places where stolen equipment and materials are discovered and recovered. NER does an excellent job of that in its report for heavy machinery. This example is a great template for the entire program.

These statistics would greatly assist law enforcement, security, construction, and insurance professionals to gain a firm understanding of the scope of the problem and would help them to find solutions, such as those in this book, to combat the problems associated with construction sites.

CHAPTER POINTS BY DISCIPLINE

1. **Security professionals:** Standards in the security industry are critical to your ability to serve your clients by deterring crime and other mishaps. After all, that is why they hired you. Your clients expect your security officers to be properly trained and licensed. They deserve that and more. Following established standards will further serve to enhance your company's reputation throughout the industry.

   The standards in this chapter are designed to make the construction site a safer, crime-free, and more profitable area in which your client can operate. There are decisions about these standards that only your client can make. However, as a security professional, you can advise your clients of these standards and make suggestions to them. You must be able to demonstrate the logic of these standards to your clients to help them make better security decisions.

   If a client balks at the suggestions, do your best to apply as many of the standards as you can without upsetting them. Your diligence will pay off in spades.

   It is of prime importance to the industry that you heed the standards set forth in the employee screening segment. The stereotype of private security officers can only be changed positively by screening and hiring of professional personnel. The securing of critical construction projects requires nothing less than full attention to employee screening.

2. **Construction professionals:** You as construction professionals are the main driving force behind establishing security standards in the industry. Without your efforts, nothing will change, and the thieves will win.

   Many of you already have strong security programs that already meet some of the standards set forth in this chapter. I ask you to consider the standards that you have not yet adopted. I realize that these are costly and time consuming.

But how costly and time consuming is dealing with a major theft at one of your construction sites?

To help further these principles, it is helpful to discuss them at meetings with other construction professionals, as well as clients whenever you have the opportunity. It is through your efforts that the culture of "rolling the dice" will go the way of the buggy whip and the VHS recorder.

3. **Law enforcement professionals:** If your duties as a law enforcement professional bring you into the realm of outdoor construction projects, then you should familiarize yourself with the standards set forth in this chapter.

    Some construction companies have their own proprietary security and might want to consult with you to evaluate their programs. If you make your rounds to the construction sites in your area as you should, you will develop the kind of relationships that will make your opinions more valuable and sought after. This approach is not totally altruistic. Helping with the security at outdoor construction sites will help you both in the reduction of crime and the swifter apprehension of criminal suspects.

4. **Insurance professionals:** Evaluate the standards set forth in this chapter and use them to make recommendations to all of your clients. The organizations that contract for the outdoor building jobs (i.e., the construction companies and the security providers) will recognize and appreciate that your recommendations pay off in the reduction of claims. If you walk your sites with your clients, you can offer suggestions to help your clients shore up their security programs and stop "rolling the dice."

5. **Legal professionals:** Whether your clients are construction professionals, security professionals, litigants who are opposing these organizations, or criminal defendants, it pays to be aware of industry standards when it comes to construction site security. It is through litigation and prosecution that many standards become legal requirements.

    Often your role as protector of your clients is overlooked. You provide the leadership and guidance that can keep your client either out of court or minimally at risk. If you present the concepts in this chapter to your client and contrast them against the cost of litigation and other losses associated with outdoor construction sites, you will be providing an invaluable service to them and perhaps avoid the costs associated with having to take a case to trial.

    As judges, you can use the standards set forth here to evaluate the reasonableness of security programs at construction sites during cases before your court. With this information, your decisions will be both knowledgeable and just. A decision that you make on such a case could establish a legal standard for the construction or security industry.

Construction Equipment 101

10

CHAPTER OUTLINE

This chapter is an appendix for the uninitiated to educate them on the various types of equipment. This includes photographs, descriptions of the equipment, what each piece does, and how much they cost.

REAL CONSTRUCTION PEOPLE, PLEASE BEAR WITH US

Okay. Stop laughing. All of the machinery in this chapter is already well known to you. But most nonconstruction professionals wouldn't know a backhoe from a Bidwell machine. The concept of this chapter is to provide a quick reference for those in the other disciplines so they can know exactly what you are talking about when you have a security issue with one of your pieces of equipment.

This section is by no means exhaustive, but it does cover most of the basic construction equipment that one is likely to encounter at any construction site, anywhere in the world.

You won't see any advertisements for any brand of equipment. There are great manufacturers all over the world, and this chapter will show some of their products. The prices range from used to new equipment. Some larger mining-sized equipment will be higher priced than I quote.

AERIAL WORK PLATFORM

This machine (Figure 10.1) lifts workers to areas that are too high to safely access with a ladder. Depending on the type of device, it can also be called a scissor lift or a cherry picker. These machines are usually small enough to be trailered away or put aboard a large truck and driven away.

Price in USD: Between $4000 and 80000.

FIGURE 10.1

Aerial work platform.

AIR COMPRESSOR

An air compressor (Figure 10.2) is a motorized machine that compresses air for use for a variety of purposes. Smaller air compressors can do simple tasks such as inflating tires, and larger ones can be used to operate in a construction environment to power a variety of pneumatic tools, such as a jackhammer or a paint sprayer.

These may vary in size from a compressor light enough to be carried away by hand to a large machine attached to a trailer that can be hauled behind a truck or other vehicle.

Price in USD: Between $500 and $55,000.

FIGURE 10.2

Air compressor.

ARROW SIGN

An arrow sign (Figure 10.3) in construction is a lighted sign, usually battery powered or partially solar powered, that displays a variety of lighted arrows that indicate to motorists the direction that they should take in order to drive around an obstacle or blocked traffic lane. Nearly all of these signs today are mounted on a trailer that can be hooked up to a vehicle and easily and quickly be towed away. Some companies take the wheels off, making the trailer more difficult to steal. Thieves still steal the batteries out of them; steal their solar panels, if so equipped; or sell the battery cables for the copper in them.

Price in USD: Between $1500 and $7000.

FIGURE 10.3

Arrow sign.

ASPHALT BATCH PLANT

This plant (Figure 10.4) is usually a temporary structure that is erected when there is a large paving project that involves the production of asphalt or macadam. It heats and mixes the roadstone with tar or bitumen for loading into dump trucks.

Price in USD: Between $50,000 and $1,250,000.

FIGURE 10.4

Asphalt batch plant.

BACKHOE

A backhoe (Figure 10.5) is a device used for excavating or digging that has a digging bucket attached to an articulated arm. It is normally attached to the back of a tractor or a front loader. It is generally used to dig trenches or ditches in dirt.

Price in USD: Between $5000 and $45,000. (Most backhoes come as a backhoe–loader combination.)

FIGURE 10.5

Backhoe.

BACKHOE LOADER

This machine (Figure 10.6) has a loader shovel attached to the front and a backhoe device attached to the back. It is versatile and can be found on construction sites throughout the world, being used to move dirt, as well as dig trenches and ditches. In Britain, it is sometimes referred to as a JCB, after its British inventor, J.C. Bamford.

Price in USD: Between $10,000 and $75,000.

FIGURE 10.6

Backhoe loader or JCB.

BIDWELL MACHINE

A Bidwell machine (bridge paver or bridge deck finisher) is a concrete road construction machine that consists of a bridge set on either end on rails and powered by a diesel engine. It is designed to roll the concrete flat and to provide a structure from which concrete workers can finish the concrete from above the roadway.

It is possible to think that no one will steal batteries from a Bidwell machine because they are often located in hard-to-reach places. Think again. Crooks will go through all kinds of contortions to make a few bucks off a stolen battery.

Price in USD: Between $5000 and $100,000.

BOBCAT

A Bobcat (Figure 10.7) is a brand of Skid Steer Loader and was invented in the United States in the 1950s. It works by the wheels on either side of the vehicle

FIGURE 10.7

Bobcat.

operating at different speeds, thus making it able to turn in a very short radius. It is a small loader, generally with a bucket attached to the front, and can be used to move dirt and a variety of material or debris. A variety of attachments can be purchased for these versatile machines, including but not limited to chippers, dump bins, bore pigs (drills), diggers, mowers, log splitters, farm equipment, utility spades, post drivers, snow blowers and snow ploughs, stump removers, and power brooms. The list is exhaustive, and one could almost devote an entire chapter to this one versatile little machine. It will perform almost any task than can be imagined if given the correct attachment.

Although many manufacturers make this type of machine, they are almost universally referred to as Bobcats because of the famous manufacturer.

A great feature that the newer Bobcat machines have is a keyless start system. This helps to combat the single key system that many manufacturers have for their fleets.

Price in USD: Between $4000 and $85,000.

BULLDOZER

A bulldozer (Figure 10.8) is a tracked tractor equipped with a large blade on the front. This earth mover is used to push large amounts of dirt, rubble, sand, or other material around construction projects. Many of them are equipped with a ripper on the back that is used to break up tightly compacted dirt or other material.

Price in USD: Between $5000 and $95,000.

FIGURE 10.8

Bulldozer.

CHERRY PICKER OR BUCKET LIFT

A cherry picker (Figure 10.9) is basically an elevated work platform upon which workers can perform their tasks high above the ground. A cherry picker can be attached to either an hydraulic lifting boom on a truck or trailer or attached to a forklift. As with many pieces of construction equipment, they originated in the agriculture business and were quickly adapted for construction. They can, in some cases, eliminate the need to construct scaffolding, which saves time and money and eliminates the risks associated with scaffolding. Typically, these machines are operated with a safety harness for each worker.

Price in USD: Between $4000 and $30,000.

FIGURE 10.9

Cherry picker.

CONCRETE BATCH PLANT

A concrete batch plant (Figure 10.10) can be either at a fixed location with the concrete trucked to the construction site or be set up as a temporary site adjacent to the construction site. Either type of site should be secured.

Temporary sites are at times set up with a concrete crusher nearby when reconstructing concrete highway. That way, the broken concrete does not have to be hauled away. It can simply be crushed and reused in the new concrete, thus saving material hauling time, as well as money spent for dumping the old concrete and for purchasing new gravel for the concrete.

Price in USD: Between $12,000 and $200,000 for a small plant. The more that is added, the more costly they get.

FIGURE 10.10

Concrete batch plant.

CONCRETE (OR CEMENT) MIXER

A concrete mixer or cement mixer (Figure 10.11) is a conical metal barrel attached generally to a trailer that turns with either gears or a chain, which is attached to a motor. This machine mixes cement and sand or gravel with water to make concrete. It saves a lot of time and labor on small jobs because it frees the workers from having to mix the concrete by hand in a tub.

Many, but not all, of these machines are small enough to either be towed away by or placed in the bed of a pickup truck, thus making them particularly attractive to thieves.

Price in USD: Between $100 and $3000.

FIGURE 10.11

Concrete mixer.

CONCRETE (OR CEMENT) MIXER TRUCK

Concrete mixer trucks (Figure 10.12) are simply cement mixers attached to a truck. Generally, these are big enough to produce a significant amount of concrete for larger projects, and are helpful in conjunction with other concrete mixers when a high volume of concrete is needed to construct a large outdoor capital project.

Price in USD: Between $30,000 and $175,000.

FIGURE 10.12

Concrete mixer truck.

DREDGE

A dredge (Figure 10.13) is an excavation device designed to remove or mine sediments or minerals from underwater locations. They can be set up on a barge or set up on shore, and they extract material from a seabed, lakebed, or riverbed using either scoops or suction hoses.

Sand and gravel for construction can be extracted from bodies of water by dredges. They present a unique security challenge because they are frequently located in areas that have easy access by the general public.

Price in USD: Between $115,000 and $1,835,000.

FIGURE 10.13

Dredge.

DUMP TRUCK

A dump truck (or tipper truck) (Figure 10.14) is a truck with an open box in the back that can be tilted up hydraulically to dump material through a tailgate. Generally, they are used at construction sites to discharge sand a gravel where needed. They vary greatly in size from small and medium trucks to large trucks as seen on highway projects to giant trucks used in mining or very large construction projects.

Price in USD: Between $3500 and $199,000.

FIGURE 10.14

Dump truck.

EXCAVATORS

An excavator (or power shovel) (Figure 10.15) is a machine used for digging large amounts of material by use of a bucket scoop attached to a mechanical or hydraulic arm. They normally sit atop a tracked or wheeled platform and have the ability to swivel on a base. The operator sits at the control station in the "house" and operates the machine from there.

Excavators vary greatly in size and appearance. Technically, a backhoe is an excavator. Some excavators used in huge projects can weigh more than 1000 tons.

Subsequently, but not necessarily in order, are listed some of the many types of excavators.

Price in USD: $6000 to $1,100,000.

FIGURE 10.15

One of many types of excavators.

BUCKET WHEEL EXCAVATOR

A bucket wheel excavator (Figure 10.16) is used primarily in mining and is designed to dig continuously. As the name implies, many buckets are attached to a giant wheel that rotates at the end of a giant cutting boom. The buckets discharge their load of material onto a conveyor attached to the cutting boom. That cutting conveyor feeds a discharge conveyor belt, which carries the material away from the machine.

Older bucket wheel excavators moved on rails, but the newer ones move on tracked crawlers. They are among the largest machines ever built by humans.

Price in USD: Between $100,000 and $3,000,000.

FIGURE 10.16

Bucket wheel excavator.

COMPACT EXCAVATOR

Compact excavators (Figure 10.17) are small excavators that generally range in size from a half ton to 8 tons. Most of them are small enough to be trailered away. They operate hydraulically and are composed of a house, an undercarriage, and a work group. Although they are generally used for digging, a variety of attachments can be placed on the end of the work group such as a grapple, an auger, or a breaker. Most are also equipped with a backfill blade that can be used as a dozer, which makes this machine particularly versatile.

Price in USD: Between $15,000 and $105,000.

FIGURE 10.17

Compact excavator.

COMPACTOR (HANDHELD)

Handheld compactors (Figure 10.18) are used to tamp down and solidify loose soil or other material and come in a variety of designs. They are small enough to be operated by hand as the name implies, and they can be easily transported away.

Price in USD: Between $2000 and $2500.

FIGURE 10.18

Hand held compactor.

COMPACT TRACK LOADER

The compact track loader (Figure 10.19) is similar in appearance to the skid steer loader, but it has tracks rather than wheels. It works by the tracks on either side of the vehicle operating at different speeds, thus making it able to turn in a very short radius. It is a small loader, generally with a bucket attached to the front, and can be used to move dirt and a variety of material or debris.

Just like the other Bobcat machines, Bobcat track loader machines have is a keyless start system. This helps to combat the single-key system that many manufacturers have for their fleets.

Price in USD: Between $16,000 and $90,000.

FIGURE 10.19

Compact track loader.

DRAGLINE EXCAVATOR

A dragline excavator's (Figure 10.20) bucket is suspended on cables from a crane boom, and when laid on the area to be excavated, it is dragged across the surface by use of a dragline cable.

Dragline excavators are used primarily in mining. They are huge in size. Most are moved by "walking" them on huge pontoons because most of them are too heavy to be effective with track propulsion. The new larger ones are very expensive, some costing between $50 and 100 million apiece.

Price in USD: Between $10,000 and $100,000,000.

FIGURE 10.20

Drag Line excavator.

LONG-REACH EXCAVATOR

A long-reach excavator (Figure 10.21) is an excavator with an extra-long boom. It was developed in England in the past century and was originally designed for excavating waterways. It is not to be confused with high-reach excavators, which are used for demolition. They are similar in appearance. They can be operated from shore or mounted on a ship or barge.

Price in USD: Between $30,000 and 1,000,000.

FIGURE 10.21

Long reach excavator.

FELLER BUNCHER

A feller buncher (Figure 10.22) is awesome to watch in operation. It was originally designed for the logging industry. It literally bunches trees together and fells them together. It can be found at large capital outdoor construction projects where trees have to be cleared to make room for the project.

Price in USD: Between $19,000 and $450,000.

FIGURE 10.22

Feller buncher.

FORKLIFT TRUCK (FORKLIFT)

A forklift truck (Figure 10.23) is most commonly referred to as just a forklift. It is designed to pick up materiel and move it short distances. In commercial use, they often move things that are stacked on wooden pallets, but pallets are not an absolute requirement. They are also used to move metal bins. There is a host of other applications for these versatile machines.

Price in USD: Between $5000 and $130,000.

FIGURE 10.23

Forklift truck.

GENERATOR

The simple definition of a generator (Figure 10.24) is a device that converts mechanical energy into electrical energy. They are used at construction sites for a variety of purposes when there is no electrical power source. They vary greatly in size, from a small machine that can be easily picked up and carried away to those that are attached to small trailers to huge models that are mounted on the ground, trucks, or large trailers. They are particularly popular item for theft because many of them are small enough to be easily stolen and are not difficult to resell.

Price in USD: Between $200 and $725,000.

FIGURE 10.24

Generator.

GRADER

A grader (Figure 10.25) (or road grader, or motor grader) is a motorized machine with a large blade designed to smooth large surfaces such as roads or runways. Most graders have two axles at the back and a third axle in front.

Price in USD: Between $50,000 and $500,000.

FIGURE 10.25

Grader.

JACKHAMMER

A jackhammer (Figure 10.26) (or demolition hammer) is actually a pneumatically operated drill that is generally powered by either an electric motor or compressed air. (Larger jackhammers can be hydraulic, but they are affixed to a rig on a larger piece of construction equipment.) They use a chisel that is propelled by a hammer and are used to break up concrete and other hard material.

Price in USD: Between $1000 and $37,000.

FIGURE 10.26

Jackhammer.

LIGHT TOWER

A light tower (Figure 10.27) is a device, usually attached to a trailer, that has a generator and a mast with metal halide lights attached and is used to illuminate construction areas. They can have two or more lights attached to the mast. These are particularly easy to steal because they are on small trailers and can be stolen in seconds if not properly secured.

Price in USD: Between $3000 and $36,000.

FIGURE 10.27

Light tower.

LOADER

A loader (Figure 10.28) is a motorized tractor with a large scoop in front that is attached to articulated arms and is used to load large quantities of material into the back of a dump truck. They can vary greatly in size from skid steer tractors to giant loaders that are used in mines and gravel pits.

Price in USD: Between $13,000 and $250,000.

FIGURE 10.28

Loader.

PAVER

A paver (Figure 10.29) is a construction machine that lays asphalt flatly on a highway, road, parking lot, school playground, bridge, or wherever such material is needed. It provides a modicum of compaction before the asphalt being compacted by a road roller.

Price in USD: Between $20,000 and $450,000.

FIGURE 10.29

Paver.

PICKUP TRUCK

A pickup truck (Figure 10.30) is a light vehicle with a cab and a bed and is perhaps the most versatile—and most stolen—vehicle in the construction industry. They have either standard cabs or crew cabs for transporting workers from place to place. Many construction workers use their own pickups at work and as personal vehicles.

Price in USD: Between $3000 and $70,000.

FIGURE 10.30

Pickup truck.

PILE DRIVER

A pile driver (Figure 10.31) is a tall structure with a framework that allows a guided weighted hammer to be dropped down atop a pile and drive it into the ground or seabed or riverbed. They vary considerable in size and design. Some are mounted on tracked or wheeled vehicles, others are mounted on platforms that lie flat on the ground or that roll along on rails, and still others are mounted on barges.

Price in USD: Between $2000 and $500,000.

FIGURE 10.31

Pile driver.

RECLAIMER

A reclaimer (Figure 10.32) is a large bridge-like structure that is laid across a roadway and travels along rails. It is used to reclaim material from a stockpile and spread it evenly on a roadway.

Price in USD: Between $85,000 and $1,000,000.

FIGURE 10.32

Reclaimer.

ROADHEADER

A roadheader (Figure 10.33) is essentially a boring machine used to excavate shafts or tunnels. Roadheaders consist of a cutting head mounted on a boom and generally have some sort of conveyer to propel the excavated material toward the rear of the machine. They can be manipulated to carve out the excavation in a variety of shapes.

Price in USD: Between $30,000 and $10,000,000.

FIGURE 10.33

Roadheader.

ROAD ROLLER

A road roller (Figure 10.34) is machine that compacts soil, asphalt, or other materials by use of a heavy roller both front and rear. In years past, they were called steam-rollers because of their propulsion by steam engines. Many people still call them steamrollers today.

Price in USD: Between $4000 and $250,000.

FIGURE 10.34

Road roller.

SKIDDER

Skidders (Figure 10.35) are large machines used mostly in the lumber industry to move logs. Occasionally, you will see them at construction sites when a forested area must be cleared to make room for the construction site.

Price in USD: Between $25,000 and $250,000.

FIGURE 10.35

Skidder.

SKID STEER LOADER

The skid steer loader (Figure 10.36) was invented in the United States in the 1950s. It works by the wheels on either side of the vehicle operating at different speeds, causing one side or the other to skid, thus making it able to turn in a very short radius. It is a small loader, generally with a bucket attached to the front, and can be used to move dirt and a variety of material or debris.

Although many manufacturers make this type of machine, they are almost universally referred to as Bobcats (see Bobcat). Again, an excellent feature that the newer Bobcat skid steer loader machines have is a keyless start system. This helps to combat the single key system that many manufacturers have for their fleets.

Price in USD: between $25,000 and $75,000.

FIGURE 10.36

Skid steer loader.

SOIL COMPACTOR

As the name implies, a soil compactor (Figure 10.37) is a machine used to compact soil or other material to make it more solid and stable to travel upon. They generally have either a spiked or smooth roller, and some are equipped with a blade to assist in smoothing out the road surface. There are also hand-operated compactors that are used to tamp down soil or other materials in small areas.

Price in USD: Between $15,000 and $550,000.

FIGURE 10.37

Soil compacter.

TELESCOPIC HANDLER

A telescopic handler (Figure 10.38) is a construction machine equipped with a long extending boom that is used to move a variety of material from place to place. Typically, they are equipped with pallet forks, but they can also be equipped with a winch or a bucket.

Price in USD: Between $9000 and $1,000,000.

FIGURE 10.38

Telescope handler.

TRACTOR

A tractor (Figure 10.39) is a machine used to haul a trailer or a variety of other types of equipment at slow speeds. They are used everywhere in the world in both construction and agriculture. They come in many designs, depending on their intended use.

Price in USD: Between $5000 and $200,000.

FIGURE 10.39

Tractor.

TRENCHER

A trencher (Figure 10.40) is a machine used in construction for digging trenches, primarily for laying pipe or cable. They vary greatly in size, depending on the nature of the job. Some are small walk-behind machines that can be operated by one operator, but others are quite large, motorized, and propelled either on wheels or on tracks.

Price in USD: Between $2000 and $565,000.

FIGURE 10.40

Trencher.

WALK-BEHIND POWER TROWEL

A walk-behind power trowel (Figure 10.41) is used to smooth concrete and is operated by one operator. It has a motor and blades that rotate. Some "walk-behind" power trowels actually have a seat and are ridden about by the operator.

Price in USD: Between $300 and $9000.

FIGURE 10.41

Walk-behind trowel.

WATER TOWER

FIGURE 10.42

Water Tower.

A portable Water Tower is a towable storage tank that is generally elevated hydraulically or by cable (Figure 10.42). They vary in size from 500 to 12,000 gallons.

Price in USD: Between $7,500 and $75,000.

CHAPTER POINTS BY DISCIPLINE

1. **Security professionals:** As a security professional, you must learn your client's business, regardless of the type of endeavor in which they are engaged. I recall a client asking me for a last-minute security officer to guard a Bidwell machine. I thought he was kidding me because I had never heard of such a thing. Luckily, I dispatched the guard and then researched just what in the world he was talking about. That is part of the reason that I decided to write this chapter, in order to provide a quick desk reference. There are only 40 or so machines covered here, but these are the most important ones. When you reference each machine, you can get an idea of the size and value of the equipment. Look at whether the equipment is small enough to be easily stolen or whether there might be valuable cables or other attachments that might be attractive to thieves.

2. **Construction professionals:** Try not to laugh at us nonprofessionals who know very little of your world. This chapter is designed to assist the people who help secure and recover your equipment by creating a ready reference for the vast array of equipment you use to build the marvels of the modern world.

3. **Law enforcement professionals:** Both patrol officers and detectives need to have an idea of what the equipment that is stolen looks like, particularly if you are broadcasting information on the theft over the radio to officers on your own department or other agencies. Even if you routinely investigate theft from construction sites, be assured that a number of your compatriots have no earthly idea what this stuff looks like and that they will be greatly assisted in the moment if they have an idea of what they are looking for.

4. **Insurance professionals:** Perhaps this reference will be most helpful when you are assisting your clients in recovering from a loss. Knowing what has been lost can help you to find a temporary equipment replacement so that your client can get back to work rapidly. Additionally, being able to identify the lost equipment will ensure that you and your client are speaking the same language.

5. **Legal professionals:** Whether you are a prosecutor, a defense attorney, or a judge, you need to know the nature of the equipment that has been stolen so you properly charge, defend, or judge a case. You should know not only the monetary value of the equipment but also how its loss impacts the victim of the crime.

   If you are a legislator who is writing new law to address cases of construction equipment theft, you should be familiar with the industry's equipment so that when people are advising you or your staff or testifying before you in committee, you understand their language.

Organizations

11

CHAPTER OUTLINE

This chapter lists a number of organizations that relate to the topics of this book. I believe that professional organizations bring value and education to their members. They give a chance for like-minded people to gather in an environment of mutual purpose to further the goals of their profession or passion. This list is by no means exhaustive but rather highlights the organizations that I know and belong to or ones that I have found in my research and interviews.

Some of the descriptors of these organizations come from their own website information or are paraphrased from that information. The intent is to provide the reader with the information as presented by each organization.

SECURITY ORGANIZATIONS
ASIAN PROFESSIONAL SECURITY ASSOCIATION

The Asian Professional Security Association has chapters in India, Thailand, the Philippines, Singapore, Malaysia, South Korea, Viet Nam, Indonesia, China, and Hong Kong. Its goal is to promote security management practices that ensure confidentiality, integrity, and availability of information resources. It also promulgates training and information in the areas of private investigations and physical security.

It has an annual conference that is held in a different Asian city each year. Each nation's chapter has its own email address. For a point of contact, I will give the India chapter's website, www.apsa-india.org.

THE AMERICAN SOCIETY OF INDUSTRIAL SECURITY INTERNATIONAL

The American Society of Industrial Security International (ASIS International) This organization is the one that I recommend all of the readers of this book join. If your professional career in any way touches the security of large or small outdoor capital projects, this organization is the premier worldwide organization wherein professionals of all works can gather to discuss and learn about industrial security and become certified in various aspects of applied security principles.

ASIS is the recognized world leader in board certifications for security professionals.

Find the closest chapter and get involved. In addition to learning about security and participating in training, you will meet a great group of people.

This worldwide organization can be contacted online at www.asisonline.org.

ASSOCIATION OF THREAT ASSESSMENT PROFESSIONALS

The Association of Threat Assessment Professionals (ATAP) is a nonprofit organization whose objective is to educate professionals about how best to protect victims of stalking, harassment, and threat situations. Its mission is to share and facilitate the experiences and techniques of professionals in the field of threat assessment and threat management. The organization serves to create both a professional and academic environment where flow of information is fostered.

> Association of Threat Assessment Professionals
> 1215 K Street, Suite 940
> Sacramento, CA 95814
> Phone: 916-231-2146
> Fax: 916-231-2141

CANADIAN ASSOCIATION OF DEFENCE AND SECURITY INDUSTRIES

The Canadian Association of Defence and Security Industries (CADSI) assists the private security industry in partnering with the military and government to assure the security of Canada. The security industry helps to protect Canada from crime, disasters, environmental emergencies, and terrorism. It puts on an annual exhibition, SECURETECH, which features conferences and a large trade show.

> The Canadian Association of Defence and Security Industries
> 300-251 Laurier Avenue West
> Ottawa, Ontario, Canada K1P 5J6
> Phone: +1-613-235-5337
> Fax: +1-613-235-0784
> Email: cadsi@defenceandsecurity.ca
> www.defenceandsecurity.ca

CANADIAN SECURITY ASSOCIATION

The Canadian Security Association (CANASA) is a national nonprofit organization dedicated to advancing the security industry and supporting security professionals in Canada. CANASA protects and promotes the interests of its members and the safety of all Canadians through education, advocacy, and leadership. CANASA advocates, educates, and provides leadership to its members in a self-regulated environment of Canadian security professionals.

> Canadian Security Association
> National Office
> 50 Acadia Avenue, Suite 201
> Markham, ON, Canada L3R 0B3
> Phone: 905-513-0622
> Toll free in Canada: 800-538-9919

Fax: 905-513-0624
Email: staff@canasa.org
www.canasa.org

CENTRAL ASSOCIATION OF PRIVATE SECURITY INDUSTRY (INDIA)

The Central Association of Private Security Industry (CAPSI) represents the leadership and workforce of the more than 7 million security officers in India. Its leadership is made up of professionals from the government and military, as well as the private sector. It has branches in each state in India and promulgates global standards and best practices in the security industry. It operates as a federal body to address and resolve security issues.

CAPSI publishes a periodical, *Security Post*, which is available online.

276, Sultan Sadan
Lane No.-3, West End Marg
Saidullajab, New Delhi 110030
Phone:- +91-11-40820070
Fax:- +91-1140820071
Email: info@capsi.in
www.capsi.in

CHINA SECURITY AND PROTECTION INDUSTRY ASSOCIATION

The China Security and Protection Industry Association (CSPIA) is the trade umbrella association for the professional security industry in the People's Republic of China. It was founded by the governing body of the Ministry of Public Security of China. Its leadership is appointed by the Ministry of Security Industry, thus giving it a strong link with government, to assist the security industry.

It puts on an annual conference, Security China, which is a trade show, exhibiting the latest in security related equipment. It publishes an annual journal, "China Security and Protection Year Book."

China Security and Protection Industry Association
Hongyang Business Building, 25, Room 203
Nanbinhe Street, Xuanwu District
Beijing, China
Phone: +86-10-51920615
Fax: +86-10-51920049
Email: International@bizcspia.com
www.bizcspia.com or www.securitychina.com.cn

INFRAGARD (USA)

InfraGard is a partnership between the FBI and the private sector. It is an association of persons who represent businesses; academic institutions; federal, state, and local law enforcement agencies; and other participants dedicated to sharing information and intelligence to prevent hostile acts against the United States.

www.infragard.org

INTERNATIONAL PROFESSIONAL SECURITY ASSOCIATION

The International Professional Security Association (IPSA) was formed to ensure professionals in the management of private security companies. Based in the United Kingdom, the organization is composed of 14 regions, including nations outside of Great Britain. It provides standardized training for both managers and security officers in many aspects of security work. It also assists members with job searches.

IPSA publishes a manual for security officers and managers titled *IPSA Security Instruction and Guidance Manual*.

International Professional Security Association
Railway House
Railway Road
Chorley, England PR6 0HW
Phone: 0845-873-8114 or +44-1257-249945
www.ipsa.org.uk

NATIONAL ASSOCIATION OF SECURITY PROFESSIONALS (UK)

The National Association of Security Professionals (NASP) assists the more than 500,000 security professionals in the education and training needed to obtain and maintain their security licenses in the United Kingdom. It provides a job search feature and provides its members with a variety of discounts on good and services.

National Association of Security Professionals Head Office
34 Woodlands Avenue
Rayleigh
Essex, England SS6 7RD
www.nasp.org.uk

SECURITY INDUSTRY ASSOCIATION (USA)

The Security Industry Association (SIA) is an organization that advocates for security policies and legislation both on a national level and in each of the 50 states. It conducts research and analysis of current security industry trends, actively develops standards, and provides training and education.

Furthermore, the SIA develops strategic alliances worldwide. It is also the sole sponsor of the ISC Expos, which contains trade shows and conferences.

It provides the education and certification for the CSPM (Certified Security Project Manager) credential as well as numerous other trainings and seminars.

The SIA also conducts an annual Government Summit that brings leaders from the security industry together with decision makers in the government.

Security Industry Association
8405 Colesville Road, Ste. 500
Silver Spring, MD 20910
Phone: 301-804-4700
Fax: 301-804-4701
www.siaonline.org

SECURITY PROVIDERS OF AUSTRALIA LTD.

This Australian association exists to improve ethics and standards in the security industry through training and development as well as monitoring and promulgating statutes and regulations for the security industry.

The Security Providers of Australia Ltd. (SPAAL) website provides a newsfeed of both national and state security–related stories.

Security Providers Association of Australia Limited
Suite G02, Macarthur Point
25 Solent Circuit
Baulkham Hills BC, NSW, Australia 2153
www.spaal.asn.au

CONSTRUCTION ORGANIZATIONS
AMERICAN CONCRETE INSTITUTE

The American Concrete Institute (ACI) is a leading authority and resource worldwide for the development and distribution of consensus-based standards, technical resources, educational programs, and proven expertise for individuals and organizations involved in concrete design, construction, and materials that share a commitment to pursuing the best use of concrete.

The ACI has its own education called the ACI University. Furthermore, the ACI offers 18 certification programs designed to form a minimum qualification for personnel employed within the concrete construction industry. It also publishes numerous technical documents, handbooks, and manuals for the concrete industry.

One of the ACI's most notable annual international events is called World of Concrete. It is the only event of its kind in the world and provides seminars, exhibits, and training for professionals in the concrete construction industry.

American Concrete Institute
38800 Country Club Dr.
Farmington Hills, MI
48331-3439 USA

Phone: 248-848-3700
Fax: 248-848-3701
www.concrete.org

AMERICAN ROAD & TRANSPORTATION BUILDERS ASSOCIATION

The American Road & Transportation Builders Association (ARTBA) is the largest national construction group in the United States. It engages in advocacy to protect the construction industry in matters such as the environment, law, and regulations. It is a source of information for transportation investment, policy, safety, economics, and other transportation industry issues. It provides education and training programs for transportation design, construction, and safety executives.

American Road & Transportation Builders Association
1219 28th Street, N.W.
Washington, DC 20007
Phone: 202-289-4434
www.artba.org

THE ASSOCIATED GENERAL CONTRACTORS OF AMERICA

The Associated General Contractors of America (AGC of America) is a leading association for the construction industry. The AGC represents more than 26,000 firms, including more than 6500 of America's leading general contractors and more than 9,000 specialty-contracting firms. More than 10,500 service providers and suppliers are also associated with AGC, all through a nationwide network of chapters.

The Associated General Contractors of America
2300 Wilson Blvd., Suite 300
Arlington, VA 22201
Phone: 703-548-3118
www.agc.org

ASIAN CONCRETE FEDERATION

This Asia-wide organization was formed to promote understanding of concrete structures and provide services to the Asian construction industry through the initiation and support of international collaborative activities. These activities focus on research and technology relating to concrete and concrete structures. It publishes and disseminates information on concrete and concrete structures through publications, conferences, symposia, workshops, and seminars. It promotes the updating and revision of concrete codes and standards on structural design, materials, construction, and maintenance.

Asian Concrete Federation
Edutivity Building, 3F
Sirindhorn International Institute of Technology

Thammasat University—Rangsit Campus
99 Moo 18, Klong Nueng, Klong Luang
Patumthani 12120, Thailand
Phone: 66-0-29869009, ext. 3408
Fax: 66-0-29869009, ext. 3410
www.asianconcretefederation.org

CANADIAN CONSTRUCTION ASSOCIATION

The Canadian Construction Association's (CCA's) vision is to build Canada with ethics, skills, and responsibility. The CCA's mission is to be the national voice for the Canadian construction industry. Members firms join the CCA through their local or provincial construction associations and are entitled to membership benefits including standard documents and construction guides, as well as updates on federal public policy and regulatory requirements.

The CCA produces several publications, most notably the magazine, *Rediscover Concrete*.

Canadian Construction Association
1900-275 Slater Street
Ottawa, ON, Canada K1P 5H9
Phone: 613-236-9455
Fax: 613-236-9526
www.cca-acc.com

CEMENT ASSOCIATION OF CANADA

The Cement Association of Canada (CAC) represents the Canadian cement industry, primarily in the manufacturing sector. It advocates for legislative and regulatory fairness with the government. It advises lawmakers on codes, standards, specifications, and best practices.

The CAC represents the Canadian cement industry, including eight companies with clinker and cement manufacturing facilities, granulators, grinding facilities, and cement terminals. Together, it "strive[s] to maintain a sustainable industry as well as promote and advance the economic, environmental and societal benefits of building with cement and concrete." A founding member of the Concrete Council of Canada, the CAC builds and maintains effective working relationships with stakeholder organizations. It advances the industry's position as a proactive partner in addressing society's sustainability challenge, and in the face of climate change, the resiliency imperative.

The CAC advocates for legislative and regulatory environments that allow a fair competitive playing field for its members at all levels of government and advises on technical matters important to the cement and concrete industries, such as codes, standards, specifications, and best practices.

www.cement.ca

CIVIL CONTRACTORS FEDERATION (AUSTRALIA)

This organization has branches in every Australian state and territory. Members construct and maintain Australia's roads, bridges, pipelines, drainage, ports, and utilities. They work in the residential and commercial construction industry in earthmoving and land development, including power, water, gas, and communications services.

www.civilcontractors.com

THE CONCRETE SOCIETY (UK)

The Concrete Society in the United Kingdom is member based and independent. It is a leading provider of information for architects, engineers, specifiers, suppliers, contractors, and other users of concrete.

It provides technical information as well as training and education for those in the industry. It also publishes *Concrete* magazine, both in printed and digital forms.

The Concrete Society
Riverside House
4 Meadows Business Park
Station Approach
Blackwater, Camberley
Surrey, England GU17 9AB
Phone: +44-0-1276-607140
Fax: +44-0-1276-607141
www.concrete.org.uk

CONCRETE SOCIETY OF SOUTHERN AFRICA

The Concrete Society of Southern Africa is a nonprofit organization that promotes excellence and innovation in the use of concrete. It also provides a forum for networking and technology transfer between members and local and international affiliates.

It produces several publications, most notable the journal *Concrete Beton*.

Concrete Society of Southern Africa
PO Box 75364
Lynnwood Ridge, 0040
South Africa
Phone: +27-12-348-5305/1319
Fax: +27-12-348-6944
Email: info@concretesociety.co.za
www.concretesociety.co.za

CONSTRUCTION EQUIPMENT ASSOCIATION (UK)

The Construction Equipment Association (CEA) is the trade association that represents the U.K. construction equipment sector and is recognized by HM Government as the voice of the industry.

Members can draw on a wide range of services designed to promote the performance of the U.K. construction equipment industry, including technical and regulatory, international business, trade exhibitions, business promotion, and lobbying.

Construction Equipment Association
1 Bickenhall Mansions
Bickenhall Street
London, England W1U 6BP
Phone: +44-0-20-8253-4502
Fax: +44-0-20-8253-4510
Email: cea@admin.co.uk
www.thecea.org.uk

CONSTRUCTION INDUSTRY CRIME PREVENTION PROGRAM (CALIFORNIA, USA)

This organization is a local California statewide association. The purpose of the CICP Program is to provide members with crime prevention tools such as employee education, site security options, and a network of members and partners working toward the same goals of crime prevention at construction sites.

The reason that I have included this local organization in this list is to underscore that crime prevention at construction sites can be undertaken at any level, particularly if there is no national agency or organization that operates with these goals in mind.

www.cicpp.org

CONCRETE INSTITUTE OF AUSTRALIA

The Concrete Institute of Australia is an independent, nonprofit organization made up of members who share an interest in concrete construction in Australia. Its mission is to promote and develop excellence in concrete technology, application, design, and construction throughout the country.

It promulgates a biennial conference. The conference is dedicated to bringing together global leaders in the concrete industry, covering all aspects of concrete design improvements, research, construction, and maintenance and repair of concrete projects.

Concrete Institute of Australia
PO Box 1227
North Sydney, NSW, Australia 2059
Phone: 02-9955-1744
Fax: 02-9966-1871
www.concreteinstitute.com.au

EUROPEAN CONCRETE PAVING ASSOCIATION

The European Concrete Paving Association (EUPAVE) is a nonprofit organization formed to promote all aspects of cement and concrete products to the European transport infrastructure and to advocate and train for road safety, fuel consumption, congestion reduction, and sustainable construction. It publishes the monthly *EUPAVE Newsletter* and conducts the EUPAVE. It is headquartered in Brussels, Belgium.

www.eupave.eu

JAPAN CONCRETE INSTITUTE

The Japan Concrete Institute is the parent body for promoting research concerning concrete. It conducts investigations and research on concrete, reinforced concrete, other types of concrete, and various materials and equipment related to concrete. Through coordination of investigations and research and dissemination of the results, it furthers research and advances in technology concerning concrete.

It publishes the *Concrete Journal* monthly and distributes it free of charge to its members. Another journal, *Concrete Research and Technology*, is published in Japanese three times per year and distributed to members.

Japan Concrete Institute
12F, Sogo Hanzomon Bldg.
7 Kojimachi1-chome, Chiyodaku
Tokyo 102-0083, Japan
Phone: +81-3-3263-1571
Fax: +81-3-3263-2115
www.jci.net.or.jp

MASTER BUILDERS AUSTRALIA

Master Builders is the Australian building and construction industry association. Its role is to promote the viewpoints and interests of the building and construction industry and to provide services to members in areas that include training, legal services, industrial relations, building codes and standards, industry economics, and international relations.

Master Builders Australia
Level 1, 16 Bentham Street
PO Box 7170
Yarralumla, ACT, Australia 2600
Phone: 02-6202-8888
Fax: 02-6202-8877
www.masterbuilders.com.au

MASTER BUILDERS SOUTH AFRICA

Master Builders South Africa is the leading national representative body in the building and construction industry in South Africa. Its primary roles are to promote the viewpoints and interests of the industry and to promote quality and standards of excellence in service to its members. It engages government and legislative bodies on national policies that affect the industry for the purpose of creating a sustainable building industry in South Africa.

It represents its members on national bodies and lobbies national government on legislative and other policy issues. It also provides a range of services to its members, which encompass training needs, legal services, labor relations, building codes and standards, and economics that affect the building industry.

Master Builders South Africa
No 1 Second Road Randjespark, Midrand
Johannesburg, South Africa
Phone: 011-205-9000
Fax: 011-315-1644
www.mbsa.org.za

NATIONAL ASSOCIATION OF WOMEN IN CONSTRUCTION

The National Association of Women in Construction (NAWIC) originally began as Women in Construction of Fort Worth, Texas, USA. The founders organized NAWIC to create a support network. Women in Construction of Fort Worth was gained its national charter in 1955 and became the National Association of Women in Construction. Today, NAWIC provides its members with opportunities for professional development, education, networking, leadership training, public service, and more. The NAWIC also has affiliates in Canada, Australia, the United Kingdom, South Africa, and New Zealand.

www.nawic.org

NATIONAL EQUIPMENT REGISTER

The National Equipment Register (NER) provides recovery and risk management services for equipment owners, buyers, sellers, insurers, manufacturers, and financial institutions in the United States. It also produces and excellent annual statistical report of theft statistics of construction and other types of machinery. (When looking at the NER's statistics, keep in mind that other industries besides construction are represented, but you can nevertheless get a clear idea of what is being stolen.)

Additionally, the NER provides training to law enforcement professionals on heavy equipment theft and recovery. It has three very important help tools: IRONwatch, IRONcheck, and HELPtech. (See Chapter 13 for a more complete description of the NER.)

National Equipment Register
545 Washington Blvd.
Jersey City, NJ 07310-1686
Phone: 201-469-2030
www.ner.net

NATIONAL EQUIPMENT REGISTER AUSTRALIA

The National Equipment Register (NER) Australia is similar to the NER in the United States and the National Plant & Equipment Register in the United Kingdom in that it provides for registration of heavy equipment, records of theft, police liaison, and tracking of marked equipment.

The unique aspect of the NER Australia's tracking feature is that it uses DataDot technology, a virtually invisible and highly efficient system of preprogrammed microdots that are applied to the equipment in various locations. This is discussed in greater detail in Chapter 13.

National Equipment Register Australia
9/19 Rodborough Road
Frenchs Forest, NSW, Australia 2086
Phone: 1300-300-829 (Australia only)
Phone: +61-2-8977-4900 (worldwide)
www.nationalequipmentregister.com.au

NATIONAL INSURANCE CRIME BUREAU

The National Insurance Crime Bureau (NICB) is a not-for-profit organization in the United States that receives support from property and casualty insurance companies and self-insured organizations. The NICB partners with insurers and law enforcement agencies to facilitate the identification, detection, and prosecution of insurance criminals. This organization additionally partners with the National Equipment Register to provide it with statistical data for its annual analysis of heavy machinery theft.

National Insurance Crime Bureau
1111 E. Touhy Ave., Ste. 400
Des Plaines, IL 60018
Phone: 800-447-6282 or 847-544-7000
Fax: 847-544-7100
Law enforcement assistance: 847.544.7002
www.ncib.org

NEW ZEALAND CONCRETE SOCIETY

The purpose of the New Zealand Concrete Society (NZCS) is to encourage a greater knowledge and understanding of all aspects of structural and architectural concrete and to support their development and use where appropriate.

It produces numerous technical publications, including an annual report. It also has an annual concrete conference in New Zealand. Furthermore, the NZCS participates in the development of appropriate standards, codes of practice, and specifications.

www.concretesociety.org.nz

THE NATIONAL PLANT AND EQUIPMENT REGISTER (UK AND EUROPE)

The National Plant and Equipment Register (TER) provides heavy equipment registration in the UK and Europe and assists law enforcement and the insurance industry with loss prevention, as well as recovery of stolen heavy equipment and machinery. It is similar in scope as the National Equipment Register in the United States. More information about the TER is presented in Chapter 13.

> The National Plant & Equipment Register
> Office 2 H/I
> Wessex House
> 40 Station Road
> Westbury, Wiltshire, England BA13 3JN
> Phone: 01225-464599
> Fax: 01225-317698
> Email: info@ter-europe.org
> www.ter-europe.org

LAW ENFORCEMENT ORGANIZATIONS
ASEANAPOL

ASEANAPOL is an international police organization whose goal is to promote cooperation among law enforcement agencies in Southeast Asia. The member countries are Brunei, Cambodia, Indonesia, Laos, Malaysia, Myanmar, Philippines, Singapore, Thailand, and Viet Nam.

> ASEANAPOL
> Level 1, Tower 2
> Royal Malaysia Police
> Headquarters
> Bukit Aman
> 50560 Kuala Lumpur, Malaysia
> Phone: +603-22668821/22
> Fax: +603-22668825
> Email: aseanapolsec@aseanapol.org
> www.aseanapol.org

ASSOCIATION OF THREAT ASSESSMENT PROFESSIONALS

The Association of Threat Assessment Professionals (ATAP) is a nonprofit organization whose objective is to educate professionals about how best to protect victims of stalking, harassment, and threat situations. Its mission is to share and facilitate the experiences and techniques of professionals in the field of threat assessment and threat management. The organization serves to create both a professional and academic environment where flow of information is fostered.

Association of Threat Assessment Professionals
1215 K Street, Suite 940
Sacramento, CA 95814
Phone: 916-231-2146
Fax: 916-231-2141

EUROPEAN GENDARMERIE FORCE

The European Gendarmerie Force (EGF) is a multinational initiative of six EU Member States: France, Italy, The Netherlands, Portugal, Romania, and Spain. It was established by treaty with the aim to strengthen international crisis management capacities and contribute to the development of the Common Security and Defense Policy.

www.eurogendfor.org

EUROPOL

Europol is the European Union's law enforcement agency whose main goal is to help achieve a safer Europe for the benefit of all EU citizens. It does this by assisting the EU's Member States in their fight against serious international crime and terrorism.

Europol officers have no direct powers of arrest but support EU law enforcement colleagues by gathering, analyzing, and disseminating information and coordinating operations. Its partners use input to prevent, detect, and investigate offenses and to track down and prosecute those who commit them. Europol experts and analysts take part in Joint Investigation Teams that help solve criminal cases on the spot in EU countries.

Visitor address:
Europol
Eisenhowerlaan 73
2517 KK, The Hague, The Netherlands
Mailing address:
Europol
PO Box 908 50
2509 LW, The Hague, The Netherlands
Phone: +31-70-302-5000
Fax: +31-70-345-5896
www.europol.europa.eu

FEDERAL CRIMINAL INVESTIGATORS ASSOCIATION (USA)

The Federal Criminal Investigators Association (FCIA) serves the needs of the American federal investigative community. The mission of FCIA is to ensure that federal law enforcement professionals have the tools and the support network to meet the challenges of future criminal investigations while becoming more community oriented.

Federal Criminal Investigators Association
5868 Mapledale Plaza, Suite 104
Woodbridge, VA 22193
Phone: 800-403-3374
www.fedcia.org

INFRAGARD (USA)

InfraGard is a partnership between the FBI and the private sector. It is an association of persons who represent businesses; academic institutions; federal, state, and local law enforcement agencies; and other participants dedicated to sharing information and intelligence to prevent hostile acts against the United States.

www.infragard.org

INTERNATIONAL ASSOCIATION OF CHIEFS OF POLICE

The International Association of Chiefs of Police (IACP) serves as the professional voice of law enforcement. The IACP addresses cutting-edge issues confronting law enforcement through advocacy, programs and research, as well as training and other professional services. The IACP is a comprehensive professional organization that supports the law enforcement leaders of today and develops the leaders of tomorrow.

The IACP's membership consists of police chiefs, commissioners, sheriffs, constables, security officers, investigators, colonels, city managers, public safety directors, instructors, highway safety specialists, police science coordinators, brigadier generals, doctors, senior research fellows, sergeants, criminal investigators, psychologists, attorneys, management analysts, border patrol agents, inspectors, human rights officers, coroners, handwriting examiners, criminal justice students, and service providers

International Association of Chiefs of Police
44 Canal Center Plaza, Suite 200
Alexandria, VA 22314
Phone: 703-836-6767 or 800-THE-IACP
www.theiacp.org

INTERPOL

Interpol is the world's largest international police organization, with 190 member countries. Interpol facilitates international police cooperation even when diplomatic relations do not exist. Its constitution prohibits any intervention or activities of a political, military, religious, or racial character.

The official name is IPCO (International Police Criminal Organization) or IPCO-INTERPOL.

> Interpol
> General Secretariat
> 200, quai Charles de Gaulle
> 69006 Lyon, France
> Fax: +33-0-4-72-44-71-63
> www.interpol.org

THE NATIONAL SHERIFFS' ASSOCIATION (USA)

The National Sheriffs' Association (NSA) is a professional association dedicated to serving the Office of Sheriff and its affiliates through police education, police training, and general law enforcement information resources. The NSA represents thousands of sheriffs, deputies, and other law enforcement professionals; public safety professionals; and concerned citizens nationwide.

The NSA serves as the center of a vast network of law enforcement information, filling requests for information daily and enabling criminal justice professionals, including police officers, sheriffs, and deputies, to locate the information and programs they need. The NSA recognizes the need to seek information from the membership, particularly the sheriff and the state sheriffs' associations, to meet the needs and concerns of individual NSA members. While working on the national level, NSA has continued to seek grass-roots guidance, ever striving to work with and for its members, clients, and citizens of the nation.

> The National Sheriffs' Association
> 1450 Duke Street
> Alexandria, VA 22314-3490
> Phone: 800-424-7827
> Fax: 703-838-5349
> www.sheriffs.org

INSURANCE ORGANIZATIONS
AMERICAN INSURANCE ASSOCIATION (USA)

The American Insurance Association (AIA) is a property-casualty insurance trade organization, representing approximately 325 insurers that write more than $127 billion in premiums each year. AIA member companies offer all types of property-casualty insurance, including personal and commercial auto insurance, commercial property, and liability coverage for small businesses, workers' compensation, homeowners' insurance, medical malpractice coverage, and product liability insurance.

American Insurance Association
2101 L Street, NW, Suite 400
Washington DC 20037
Phone: 202-828-7100
Fax: 202-93-1219
www.aiadc.org

AMERICAN SOCIETY OF APPRAISERS (USA)

The American Society of Appraisers (ASA) is an international organization devoted to the appraisal profession. ASA is the oldest and only major appraisal organization designating members in all appraisal specialties.

American Society of Appraisers
11107 Sunset Hills Rd, Suite 310
Reston, VA 20190
Phone: 800-ASA-VALU or 703-478-2228

GESAMTVERBAND DER DEUTSCHEN VERSICHEUNGSWIRTSCHAFT (GERMANY)

This organization represents insurance concerns in Germany.

Gesamtverband der Deutschen Versicherungswirtschaft e.V.
Friedrichstraße 191
10117 Berlin, Germany
Phone: 030-2020-5000
Fax: 030-2020-6000
www.gdv.de

ASSOCIATION OF BRITISH INSURERS (UK)

The Association of British Insurers (ABI) represents U.K. insurance companies. It provides consumers with general information on insurance and savings products and services. It promotes best practice, transparency, and high standards within the industry. The ABI works with government regulators and policymakers in the United Kingdom and internationally to ensure the insurance industry meets the highest standards. It is not a regulatory agency.

Association of British Insurers
51 Gresham Street
London, England EC2V 7HQ
www.abi.org.uk

ASSOCIAZIONE NAZIONALE FRA LE IMPRESE ASSICURATRICI (ITALY)

The Associazione Nazionale fra le Imprese Assicuratrici (ANIA) represents insurance concerns in Italy.

> Associazione Nazionale fra le Imprese Assicuratrici
> Via di San Nicola da Tolentino, 72
> 00187 Roma, Italy
> Phone: 06-326881
> Fax: 06-3227135
> www.ania.it

FÉDÉRATION FRANÇAISE DES SOCIÉTÉS D'ASSURANCES (FRANCE)

The Fédération Française des Sociétés d'Assurances (FFSA) represents insurance concerns in France.

> Fédération Française des Sociétés d'Assurances
> 26, boulevard Haussman
> 75311—Paris Cedex 09, France
> Phone: 01-42-47-93-24
> www.ffsa.fr

THE GENERAL INSURANCE ASSOCIATION OF JAPAN

The objective of the General Insurance Association of Japan (GIAJ) is to promote sound development and maintain reliability of the general insurance business in Japan and thus contribute to a secure and safe society.

> General Insurance Association of Japan
> General Insurance Building
> 9, Kanda Awajicho 2-chome
> Chiyoda-Ku, Tokyo 101-8335, Japan
> Phone: +81-3-3255-1439
> Fax: +81-3-3255-1234
> www.sonpo.org.jp

INSURANCE BUREAU OF CANADA

The Insurance Bureau of Canada's (IBC's) member companies represent 90% of the Canadian property and casualty insurance market. The IBC works on a number of fronts to increase public understanding of home, auto, and business insurance.

> Insurance Bureau of Canada
> 777 Bay Street, Suite 2400
> PO Box 121

Toronto, ON, Canada M5G 2C8
Phone: 416-362-2031
Fax: 416-361-5952
www.ibc.ca

INSURANCE COUNCIL OF AUSTRALIA

The Insurance Council of Australia (ICA) is the representative body of the general insurance industry in Australia. Its members represent more than 90% of total premium income written by private sector general insurers. ICA members, both insurers and reinsurers, are a significant part of the Australian financial services system.

Insurance Council of Australia
PO Box R1832
Royal Exchange
Sydney, NSW, Australia 1225
www.insurancecouncil.com.au

INSURANCE IRELAND

Insurance Ireland actively promotes high standards in the insurance industry in Ireland. Insurance Ireland represents 95% of the domestic market and more than 80% of Ireland's international life insurance market in Ireland.

Insurance Ireland
Insurance House
39 Molesworth Street
Dublin 2, Ireland
Phone: 01-676-1914
Fax: 01-676-1943
www.insuranceireland.eu

INSURANCE REGULATORY AND DEVELOPMENT OF INDIA

The Insurance Regulatory and Development of India (IRDAI), a government organization, represents insurance concerns in India.

Insurance Regulatory and Development Authority
3rd Floor, Parisrama Bhavan, Basheer Bagh
Hyderabad 500 004
Andhra Pradesh, India
Phone: 040-23381100
Fax: 040-6682-3334

NATIONAL ASSOCIATION OF INSURANCE COMMISSIONERS (USA)

The National Association of Insurance Commissioners (NAIC) is the U.S. standard-setting and regulatory support organization, created and governed by the chief insurance regulators from the 50 states, the District of Columbia, and five U.S. territories. Through the NAIC, state insurance regulators establish standards and best practices, conduct peer review, and coordinate their regulatory oversight. NAIC staff supports these efforts and represents the collective views of state and territory regulators, domestically and internationally. NAIC members, together with the central resources of the NAIC, form the national system of state-based insurance regulation in the United States.

National Association of Insurance Commissioners
1100 Walnut Street, Suite 1500
Kansas City, MO 64106-2197
Phone: 816-842-3600
Fax: 816-783-8175

NATIONAL INSURANCE CRIME BUREAU (USA)

The National Insurance Crime Bureau (NICB) is a not-for-profit organization in the United States that receives support from property and casualty insurance companies and self-insured organizations. The NICB partners with insurers and law enforcement agencies to facilitate the identification, detection, and prosecution of insurance criminals. This organization additionally partners with the National Equipment Register to provide it with statistical data for its annual analysis of heavy machinery theft.

National Insurance Crime Bureau
1111 E. Touhy Ave., Ste. 400
Des Plaines, IL 60018
Phone: 800-447-6282 or 847-544-7000
Fax: 847-544-7100
Law enforcement assistance: 847-544-7002
www.ncib.org

NEW ZEALAND INSURANCE LAW ASSOCIATION

The Zealand Insurance Law Association (NZILA) promotes interest in, and understanding of, the law relating to insurance and to encourage the exchange of information and ideas concerning insurance law.

The NZILA encourages collaboration between those working in the insurance industry and lawyers practicing in that field.

Its primary function is arranging seminars and conferences for topics of interest to its members. In conjunction with the Australian Insurance Law Association, a newsletter is published regularly to members.

New Zealand Insurance Law Association
PO Box 999
Auckland, New Zealand
www.nzila.org

SOUTH AFRICA INSURANCE ASSOCIATION

The South Africa Insurance Association (SAIA) represents insurance companies in South Africa and is authorized to negotiate on their behalf. The SAIA was instrumental in establishing the South African Insurance Crime Bureau (SAICB). The SAICB is an independent organization with its own members and close links to the SAIA.

The SAIA also supports the South African Police Services (SAPS) in combating crime through participation in various Business Against Crime South Africa (BACSA) initiatives. The SAIA donates a significant amount of money to BACSA on an annual basis to further specifically identified relevant initiatives.

South Africa Insurance Association
Ground Floor, Willowbrook House, Constantia Office Park
C/O 14th Avenue and Hendrik Potgieter Street, Weltevreden Park
South Africa
Phone: +27-11-726-5381
Fax: +27-11-726-5351
www.saia.co.za

SOUTH AFRICA INSURANCE CRIME BUREAU

The South African Insurance Crime Bureau (SAICB) is a nonprofit company dedicated to fighting organized insurance crimes and fraud. The SAICB has made a significant impact on insurance crime by bringing together the collective resources of insurance companies, law enforcement agencies, and other stakeholders to facilitate the detection, prevention, and mitigation of insurance crimes, as well as assist in the prosecution of repeat offenders and fraudsters through ongoing insurance fraud investigation.

South African Insurance Crime Bureau
PO Box 2522, Halfway House, 1685
South Africa
Phone: +27-11-021-1432/3/4/5
Fax: +27-86-631-7796
www.saicb.co.za

LEGAL ORGANIZATIONS
ASSOCIATED GENERAL CONTRACTORS OF AMERICA OF CALIFORNIA LEGAL ADVISORY COMMITTEE (CALIFORNIA, USA)

Attorney Ron B. Pierce of Brea, California let me know about this committee of the Associated General Contractors of America (AGC) in California. He also told me that the American Bar Association has a construction law branch. He also informed me that California construction law is different than other states in the United States. That would lead me to believe that each country in the world most likely would have unique construction laws.

Associated General Contractors of America of California
3095 Beacon Blvd.
West Sacramento, CA 95691
Phone: 916-371-2422
Fax: 916-371-2352
www.agc-ca.org

AMERICAN BAR ASSOCIATION (USA)

The American Bar Association (ABA) has nearly 400,000 members and more than 3500 entities. It is committed to serving its members, improving the legal profession, eliminating bias and enhancing diversity, and advancing the rule of law throughout the United States and around the world.

The ABA is committed to supporting the legal profession with practical resources for legal professionals while improving the administration of justice, accrediting law schools, establishing model ethical codes, and more. Membership is open to lawyers, law students, and others interested in the law and the legal profession. Each year the ABA authors and publishes more than 1000 books, periodicals, and newsletters on a wide variety of legal issues.

American Bar Association
321 North Clark Street
Chicago, IL 60654
312-988-5000
www.americanbar.org

AUSTRALIAN BAR ASSOCIATION

Since its establishment, the Australian Bar Association (ABA) has represented Australia's barristers in seeking legal reform and in representing barristers' interests and unique viewpoint in the legal profession, both in Australia and internationally. The ABA emphasizes that its members occupy a unique role in the administration of justice and the maintenance of the rule of law.

Australian Bar Association
Selborne Chambers

Lower Ground Floor
174 Phillip Street
Sydney, NSW, Australia 2000
Phone: +61-2-9232-4055
Fax: +61-2-9221-1149
Email: mail@austbar.asn.au
www.austbar.asn.au

THE CANADIAN BAR ASSOCIATION

The Canadian Bar Association (CBA) is represents all members of the legal profession. It is the voice for all members of the profession, and its primary purpose is to serve its members and provide of personal and professional development and support to all members of the legal profession. It promotes fair justice systems, facilitates effective law reform, promotes equality in the legal profession, and is devoted to the elimination of discrimination. The CBA is committed to enhancing the professional and commercial interests of a diverse membership and to protecting the independence of the judiciary and the Bar.

The Canadian Bar Association
500-865 Carling Avenue
Ottawa, Ontario, Canada K1S 5S8
Phone: 613-237-2925 or 613-237-1988
Toll free: 1-800-267-8860
Fax: 613-237-0185
www.cba.org

THE CHINA LAW SOCIETY

The China Law Society (CLS) is a national association of legal scholars, jurists, law practitioners, and an academic body of legal sciences. CLS pursues objectives of enriching legal studies, promoting rule of law and making prosperous material, political, and spiritual civilizations. It has been recognized in the Chinese society as a force for development of socialist democracy, rule of law, and promotion of human rights causes.

The China Law Society
No63, Bing Ma Si Hutong
Xicheng District
Beijing 100034, People's Republic of China
www.chinalawsociety.com

THE HONG KONG BAR ASSOCIATION

The Hong Kong Bar Association is the professional organization of barristers in Hong Kong. The objects of the Hong Kong Bar Association are to consider and to take proper action on all matters affecting the legal profession and the administration of justice. These include the maintenance of honor, an independence of the bar, the improvement of the administration of justice, and the prescribing of rules of professional conduct.

The Hong Kong Bar Association
LG2, High Court
38 Queensway, Hong Kong
Phone: 852-2869-0210
Fax: 852-2869-0189
www.hkba.org

INDIAN NATIONAL BAR ASSOCIATION

The Indian National Bar Association (INBA) represents the interest of Indian legal community and strives to provide economic and social benefits that should accrue to it. It seeks to reform the Indian legal systems to effect quick justice for everyone and to reform the Indian government's bureaucratic rules and regulations.

First Floor, Jungpura Extension
New Delhi-110014
www.indianbarassociation.org

JAPAN FEDERATION OF BAR ASSOCIATIONS

The Japan Federation of Bar Associations (JFBA) is a source of protection of fundamental human rights and of realization of social justice to maintain the role of attorneys and provides oversight of attorneys in matters relating to the guidance, liaison, and supervision.

Japan Federation of Bar Associations
1-1-3 Kasumigaseki, Chiyoda-ku
Tokyo 100-0013, Japan
Phone: +81-0-3-3580-9741
Fax: +81-0-3-3580-9840
www.nichibenren.or.jp

THE LAW SOCIETY (ENGLAND AND WALES)

The Law Society represents solicitors in England and Wales. From negotiating with and lobbying the profession's regulators, government, and others to offering training and advice, it helps protect and promote solicitors across England and Wales. The Law Society Group includes the Solicitors Regulation Authority.

www.lawsociety.org.uk

THE LAW SOCIETY OF HONG KONG

The Law Society is a professional association for solicitors in Hong Kong. The purpose of the Law Society are to support and protect the character, status, and interests of solicitors in Hong Kong; to promote good standards of practice and maintain ethical practice; to ensure compliance by solicitors with relevant laws, codes, regulations, and practice directions; to develop and maintain the work of solicitors in all areas of the law, legal practice, and procedures; to ensure the view of solicitors is accurately and purposefully communicated; to provide services to its members and to consider all questions affecting the interests of the profession; and to represent the profession to procure changes of law or practice.

The Law Society of Hong Kong
3/F, Wing On House
71 Des Voeux Road Central, Hong Kong
Phone 852-2846-0500
Fax: 852-2845-0387
www.hklawsoc.org.hk

THE LAW SOCIETY OF IRELAND

The Law Society of Ireland exercises statutory functions under the Solicitors Acts 1954 to 2013 in relation to the education, admission, enrolment, discipline, and regulation of the solicitors' profession. It is the professional body for its solicitor members, to whom it also provides services and support.

The Law Society of Ireland
Blackhall Place, Dublin 7
DX 79 Dublin, Ireland
Phone: +353-1-672-4800
Fax +353-1-672-4801
www.lawsocitey.ie

THE LAW SOCIETY OF SCOTLAND

The Law Society of Scotland is the professional body for Scottish solicitors. It regulates and represents all practicing solicitors in Scotland. All practicing Scottish solicitors are members of the Law Society and are required to meet its standards.

The Law Society of Scotland
26 Drumsheugh Gardens
Edinburgh, Scotland EH3 7YR
Phone: +44-0-131-226-7411
Textphone: +44-0-131-476-8359
Fax: +44-0-131-225-2934
www.lawscot.org.uk

THE LAW SOCIETY OF SOUTH AFRICA

The Law Society of South Africa (LSSA) promotes the substantive transformation of the legal profession through its leadership role, represents and promotes the common interests of the profession, and empowers the profession by providing training to candidate attorneys and continuing professional development to attorneys.

The Law Society of South Africa
304 Brooks Street, Menlo Park, Pretoria
PO Box 36626, Menlo Park 0102
Docex 82, Pretoria, South Africa
Phone: +27-0-12-366-8800
Fax: +27-0-12-362-0969
www.lssa.org.za

NATIONAL BAR COUNCIL OF SOUTH AFRICA

The National Bar Council of South Africa (NBCSA; formerly known as the IAASA), is a voluntary association and was formed to promote the enhancement of the profession by allowing advocates to accept briefs directly from the public. It encourages healthy competition among lawyers, including advocates and attorneys, which will translate into a better and more cost-effective service to the public. Assistance is given to previously disadvantaged individuals to enter the profession without having undue barriers of entry placed in their way.

National Bar Council of South Africa
1st Floor, Liberty Building, 21 Aurora Drive, Umhlanga Ridge
Durban
Kwa-Zulu Natal
South Africa
4000
Phone: 031-535-7108
Fax: 086-517-4789
Email: info@nationalbarcouncil.co.za
www.nationalbarcouncil.co.za

NEW ZEALAND BAR ASSOCIATION

The New Zealand Bar Association (NZBA) promotes and encourages a strong and independent bar and promotes the interests of barristers and the separate independent bar. It promotes and encourage a high standard of ethical conduct among barristers.

New Zealand Barr Association
PO Box 631
Shortland Street
Auckland, New Zealand
www.nzbar.org.nz

NEW ZEALAND LAW SOCIETY

The New Zealand Law Society provides a wide variety of services for lawyers of all disciplines in New Zealand.

New Zealand Law Society
PO Box 5041, Wellington 6145
DX SP20202, New Zealand
National office: +64-4-472-7837
Lawyers complaints service: 0800-261-801
Registry: 0800-22-30-30 (or +64-4-472-7837 if outside New Zealand)
Fax: +64-4-473-7909

Checklists*

12

CHAPTER OUTLINE

As I mentioned early in the book, I am a list maker. That's just the way what's left of my Irish brain works. I find that if I commit a plan to writing and make a list, I am much less likely to forget something or omit an important detail.

This chapter contains some worksheets for a variety of concepts covered in this book. Feel free to copy them or download them and use them at work. Having a completed checklist can further be proof of due diligence. Some of you might want to use digital photography to further document elements that have been covered during the setup of your plan. You can never have enough photographs.

---

*Checklists provided in this chapter are available at the book's companion site: http://booksite.elsevier.com/9780128023839.

SITE INSPECTION CHECKLIST

This checklist is a guide for security managers and field supervisors to use when conducting inspections of construction site security operations. Not only does it serve as a reminder of what to look for, but it also helps you note potential security and safety hazards, as well as conduct and deportment of security officers. This form can be used for both contract and proprietary security operations.

 # Site Inspection Checklist

Site _____

Date _____ Time of Inspection _____

  1. **Security Officer(s) on Duty** _____

Uniform Appearance Good ___ Competent ___ Fair ___ Needs Improvement ___
Comments _____
Attentiveness Good ___ Competent ___ Fair___ Needs Improvement ___
Comments
_____
_____
_____

  2. **Safety Equipment**

Helmet   Yes ___ No ___ N/A ___
Flashlight   Yes ___ No ___ N/A ___
Firearm   Yes ___ No ___ N/A ___ **Serviceable?** Yes ___ No ___ N/A ___
Baton   Yes ___ No ___ N/A ___
Pepper Spray/Tear Gas Yes ___ No ___ N/A ___
Handcuffs   Yes ___ No ___ N/A ___
Licenses   Yes ___ No ___ N/A ___ **Expiration Date** _____

  3. **Vehicle** (If no vehicle is on post, leave this section blank.)

Clean?   Yes ___ No ___
Operational?   Yes ___ No ___
Lights in Order?   Yes ___ No ___
Brakes in Order?   Yes ___ No ___
Dents and/or Scratches Yes ___ No ___
Tires Good ___ Fair ___ Need Replacement ___

  4. **Site**

In Order as planned?   Yes ___ No ___
Comments _____
Assets in Order?   Yes ___ No ___
Comments _____
Lighting in Order?   Yes ___ No ___ N/A ___
Comments _____
Locks and Gates Secure? Yes ___ No ___
Cameras in Order?   Yes ___ No ___ N/A ___
OTHER COMMENTS _____

Securing the Outdoor Construction Site—Kevin Wright Carney—Ref: Chapter 1.

VEHICLE INSPECTION CHECKLIST

Company vehicles are some of the most costly pieces of equipment in any company's inventory. They are a vital component of most outdoor construction sites, yet they can be the proximate cause of endless civil liability if not in good repair and if not operated in a proper manner.

Most times at secured construction sites, the security vehicle will be rotated among a variety of divers. It is rare for anyone other than a manager to have a vehicle personally assigned to him or her. That being said, it is important to have a record of a vehicle inspection for every security officer and other employee for every shift. Why? Because frequently when an employee damages a vehicle or when a vehicle becomes inoperable, the employee will not report the condition to management.

Therefore, you need a paper (or electronic) trail of that vehicle's use for each and every shift so that there is a record of damage or disrepair. If an employee does not complete the report, there is at least a chance that that employee is the one who is not exercising care when operating company vehicles. Of course, if no one looks at the vehicle inspection checklists, then the policy of inspection is meaningless. Remember that you get what you *in*spect, not what you *ex*pect.

Vehicle Inspection Checklist

Vehicle Make _____ Vehicle Number _____

Driver _____ Employee Number _____

Date _____ Time On Duty _____ Time Off Duty _____

Starting Mileage _ _ _____ Ending Mileage _____

Clean?   Yes ___ No ___

Operational?   Yes ___ No ___

Lights in Order?   Yes ___ No ___

Safety Lights in Order   Yes ___ No ___

Brakes in Order?   Yes ___ No ___

Dents and/or Scratches Yes ___ No ___

Explain Damage _____

_____

Tires   Good ___ Fair ___ Need Replacement ___

Fuel   Full ___ Half Tank ___ Empty ___

Fuel Filled This Shift?   Yes ___ No ___ Gallons/Liters of Fuel _____

Vehicle Photo'd?   Yes ___ No ___

Securing the Outdoor Construction Site—Kevin Wright Carney—Ref: Chapter 2.

MATERIALS CHECKLIST

Beyond the concern for theft, there is a need to account for your materials at the end of each shift. How much material is lost because it just "walked away" and no one noticed it? Imagine the dollars, pounds, or Euros you could save if you simply account for your materials daily.

Some construction professionals assume that after materials are delivered, they will stay put until used. Then they are surprised when a construction phase comes close to completion and there is a shortage of materials. This can be mitigated by accounting for materials at the end of the workday and comparing the data with that of the previous workday. If you have an idea of how much material should have been used, then you can discern if materials are missing.

I realize that using this form and accounting procedure will require time and effort. That is why I suggest that an employee be assigned to this task, perhaps in concert with other accounting or security duties. Perhaps the employee who is marshalling equipment and materials at the end of each shift could complete the forms. If you weigh the cost of even one theft of materials, you will see that the expenditure for this type of work is well worth the money.

I suggest that you use one checklist for each type of material stored at each location at your site where you have materials stored. This is to be used in addition to your daily materials checkout log. Electronic forms are proper as long as you have a way to store them and refer back to their information.

Materials Checklist

Location _____

Date _____ Time _____

Reporting Worker _____ Employee Number _____

Type of Material _____

Amount at Beginning of Shift _____

New Material Delivered During Shift _____

Amount Used During Shift _____

Amount Returned at End of Shift _____

Discrepancies?   Yes ___ No ___

Monetary Value of Discrepancy _____

Explanation

_____
_____

Electronic Security Checkpoints? Yes ___ No ___

Lighted? Yes ___ No ___ N/A ___   Lights Working? Yes ___ No ___ N/A ___

Cameras? Yes ___ No ___ N/A ___   Cameras Working? Yes ___ No ___ N/A ___

Fenced?   Yes ___ No ___   Fence Secured/Locked? Yes ___ No ___ N/A ___

Securing the Outdoor Construction Site—Kevin Wright Carney—Ref: Chapter 2.

SITE SECURITY PLAN

As previously stated, failing to plan is planning to fail. Before any construction project is begun, there are detailed drawings, plans, and calculations to make sure that the project is structurally sound and will stand for many years. Engineers, construction professionals, and other planners come together to make sure that no detail is left undone. This is as it should be because the lives of everyone who uses or travels on it depend on meticulous planning.

Proper security is not nearly as intricate as building large capital projects, yet it still requires detailed advanced planning. The likelihood of someone stealing a completed highway or bridge is virtually nil, but the threat that someone might steal the component materials and heavy machinery used to build the structure is great. It happens every day.

A previously discussed, the costs of these thefts can be astronomical. So the question becomes, why would you not plan the security of an outdoor capital project in detail?

After some initial site assessments by security professionals, a security planning meeting should take place between the security managers and project site managers to discuss all possible scenarios that might impact the smooth progress of the construction project plan.

Then, depending on whether the security is proprietary or through a contracted security provider, a detailed site security plan should be made and presented to the same participants of the original security planning meeting.

Of course, no plan is of value if it is not followed. There should be documented inspections to verify that the plan is being followed as designed. These inspections should also determine if revisions, adjustments, or additions need to be made to the plan. If it is decided by the responsible parties that changes are in order, then the substance of the changes should be communicated to all participants in the construction project. This will help to avoid problems when changes are noticed.

The site security plan should be distributed on a need-to-know basis to construction site managers and all security personnel involved. Sharing the specifics of the plan should not be discussed outside that circle of participants to keep the details from falling into the wrong hands.

Site Security Plan

Project _____ Job Number _____

Construction Company _____

Location _____

Project Manager _____

Telephone _____ Email Address _____

Fax _____

Security Manager _____

Telephone_____ Email Address _____

Fax _____

Security Company _____

Start Date _____ Estimated Date of Completion _____

Law Enforcement Agency _____

Telephone _____ Email Address _____

Fax _____

Fire Department Agency _____

Telephone _____ Email Address _____

Fax _____

Security Planner _____

Telephone _____ Email Address _____

Fax _____

Fax _____

Securing the Outdoor Construction Site—Kevin Wright Carney—Ref: Chapter 3, 7, and 9.

(Page 1 of 6)

Site Security Plan

Location of Site Office _____

Description of Project _____

Locations of Protectable Groupings of Machinery and Materials:

1. _____

2. _____

3. _____

4. _____

5. _____

6. _____

7. _____

8. _____

9. _____

10. _____

Securing the Outdoor Construction Site—Kevin Wright Carney—Ref: Chapters 3, 7, and 9.

(Page 2 of 6)

Site Security Plan

Number of Security Officers Required _____

Shifts:

Day _____ **hrs. to** _____ **hrs.**

P.M. _____ **hrs. to** _____ **hrs.**

E.M. _____ **hrs. to** _____ **hrs.**

Armed _____ **Unarmed** _____

Special Equipment Needed _____

Special Skills Needed _____

Vehicles Needed _____

Communications Equipment _____

Camera Equipment _____

Recordable Patrol Points _____

Signage _____

Security Officer Posts:

    1. _____

    2. _____

    3. _____

    4. _____

    5. _____

Security Officer Routes and Frequency of Checks:

    1. _____

    2. _____

    3. _____

    4. _____

    5. _____

Securing the Outdoor Construction Site—Kevin Wright Carney—Ref: Chapters 3, 7, and 9.

(Page 3 of 6)

Site Security Plan

Fenced Areas:

1. Site Office: Fenced Yes ___ No ___ Locked/Lockable Yes ___ No ___ Fence Type _____

2. Equipment Yard 1: Fenced Yes ___ No ___ Locked/Lockable Yes ___ No ___ Fence Type _____

3. Equipment Yard 2: Fenced Yes ___ No ___ Locked/Lockable Yes ___ No ___ Fence Type _____

4. Materials Yard 1: Fenced Yes ___ No ___ Locked/Lockable Yes ___ No ___ Fence Type _____

5. Materials Yard 1: Fenced Yes ___ No ___ Locked/Lockable Yes ___ No ___ Fence Type _____

6. Site Perimeter: Fenced Yes ___ No ___ Locked/Lockable Yes ___ No ___ Fence Type _____

Lighted Areas:

1. Site Office: Lighted Yes ___ No ___ Type of Lighting _____

2. Equipment Yard 1: Fenced Yes ___ No ___ Type of Lighting _____

3. Equipment Yard 2: Fenced Yes ___ No ___ Type of Lighting _____

4. Materials Yard 1: Fenced Yes ___ No ___ Type of Lighting _____

5. Materials Yard 1: Fenced Yes ___ No ___ Type of Lighting _____

6. Site Perimeter: Fenced Yes ___ No ___ Type of Lighting _____

Traffic:

1. Is site immediately adjacent to a road or highway?   Yes ___ No ___

2. Are there significant potential safety hazards from adjacent traffic?   Yes ___ No ___

3. Is there a risk that construction equipment will impede traffic?   Yes ___ No ___

4. Are there traffic-stopping protective barriers planned?   Yes ___ No ___

5. Are ingress and egress points clearly marked?   Yes ___ No ___

6. Are security personnel trained in traffic control?   Yes ___ No ___

7. Are construction site personnel trained in traffic control?   Yes ___ No ___

8. Is night lighting available for visibility and control?   Yes ___ No ___

Plan Narrative:

_____
_____
_____
_____
_____
_____
_____
_____
_____
_____

Securing the Outdoor Construction Site—Kevin Wright Carney—Ref: Chapters 3, 7, and 9.

(Page 4 of 6)

Site Security Plan

Cameras:

Reminder: The use of dummy cameras is ill advised and are not part of this plan.

1. Cameras on site?   Yes ___ No ___

2. Are cameras actively monitored?   Yes ___ No ___

3. Are camera images viewable off-site?   Yes ___ No ___

4. Are camera images recorded?   Yes ___ No ___

5. Are night cameras supported by lighting? Yes ___ No ___

6. Are camera warning signs posted?   Yes ___ No ___

Camera Locations: (Use additional sheets if necessary.)

1._____

2._____

3._____

4._____

5._____

6._____

Securing the Outdoor Construction Site—Kevin Wright Carney—Ref: Chapters 3, 7, and 9.

(Page 5 of 6)

Site Security Plan

Narrative: (Use additional sheets if necessary.)

_____
_____
_____
_____
_____
_____
_____
_____
_____
_____
_____
_____
_____
_____
_____
_____
_____
_____
_____
_____
_____
_____
_____

Securing the Outdoor Construction Site—Kevin Wright Carney—Ref: Chapters 3, 7, and 9.

(Page 6 of 6)

CONSTRUCTION PROPOSAL SECURITY ADDENDUM

This addendum is an optional form for construction professionals to use in their job proposals. With this, you set the stage that security is more than just "part of doing business" but rather a significant component of the project, for which the buyer should provide payment. One of the prime intentions of this book is to change the mindset that security is a frivolous expense about which only insurance companies care.

Whether you use this form as a calculation sheet or an actual addendum to the bid, it is important to add these figures to the overall price of the project.

It is perfectly acceptable to indicate "Included" in the segments of this form, (e.g., cameras and monitors) wherein you already own the equipment. It lets the buyer know that you are only charging them for what is needed.

Construction Proposal Security Addendum

Project _____ Job Number _____

Construction Company _____

Location _____

Project Manager _____

Telephone _____ Email Address _____
Fax _____

Security Company _____

Start Date _____ Estimated Date of Completion _____

Security Costs Breakdown:

Security Officers: _____ hours @ $_____ per hour $_____

Camera Equipment and Monitors $_____

Signage $_____

Security Vehicle(s) Usage   (Fuel and Maintenance) $_____

Security Lighting $_____

Total $_____

Securing the Outdoor Construction Site—Kevin Wright Carney—Ref: Chapter 1.

SERIOUS INCIDENT REPORT

The Serious Incident Report (SIR) is a standard report in the security industry. It provides a record of incidents that happen that go beyond the normal day-to-day mishaps. Any law enforcement officer, insurance professional, or lawyer will affirm the value of documenting serious incidents. As a construction professional, you should document as a part of your project any incident that results in injury or death and any incident that could possibly lead to civil liability.

You might counter with the argument that law enforcement or the fire department will document any serious situation that happens on your site. That may or may not be true. Many government agencies have "short form" reports that only contain a minimum of information for "routine" reports. By the time a serious incident ends up in court, the reporting officer may have handled hundreds more incidents of a variety of natures, and the details of your incident are not clear in his or her mind, beyond what is written in the report. This is why it is important to provide your own documentation.

It is perfectly acceptable (and likely desirable) to task your security provider with completing these reports. You should, as a matter of course, review and approve these reports because you may have to testify in court to their accuracy if the incident results in litigation or prosecution.

In your site file cabinet, you should have a special folder for these reports. Always make sure to include photographs in that file.

Serious Incident Report

Project _____Job Number _____

Construction Company _____

Location _____

Project Manager _____

Telephone _____ Email Address _____

Fax _____

Security Officer _____ Employee Number _____

Date _____ Time of Occurrence _____

Law Enforcement Notified? Yes ___ No ___

Fire Department Notified? Yes ___ No ___

Law Enforcement Agency _____

Report Number _____

Telephone _____ Email Address _____

Fax _____

Fire Department Agency _____

Telephone _____ Email Address _____

Fax _____

Type of Incident

Crime (Check all applicable categories)

Robbery ___ Theft___ Assault___ Battery___ Sex Crime ___ Murder___ Arson___ Other ___

Accident (Check all applicable categories)

Fatal? Yes ___ No ___ Injuries? Yes ___ No ___ Hospital transport? Yes ___ No ___

Auto/Truck Collision ___ Machinery vs Vehicle ___ Machinery vs. Object ___ Machinery vs. Person ___

Machinery Failure/Collapse ___ Person(s) Only ___ Structural Collapse ___ Drowning___

Aircraft Crash ___ Boat Accident____ Hazardous Materials ___ Electrocution ___

Fire (Check all applicable categories)

Arson? Yes ___ No ___ Injuries? Yes ___ No ___ Fatal? Yes ___ No ___ Hospital Transport? Yes ___ No ___

Vehicle ___ Machinery ___ Structure___ Materials___ Equipment___ Fuel___

Fire Department Engine Company_____ Captain_____ Time_____

Securing the Outdoor Construction Site—Kevin Wright Carney—Ref: Chapter 4.

(Page 1 of 2)

Serious Incident Report

<u>Natural Disaster</u> (Check all applicable categories)

Fatal? Yes ___ No ___ Injuries? Yes ___ No ___ Hospital transport? ___
Yes ___ No ___

Forrest Fire___ Flood____ Earthquake___ Storm___ High Winds___ Lightning___
Volcano ___ Other ___
(Explain)_____

Fire Department Engine Company_____ Captain_____
Time:_____

Narrative_____
_____
_____
_____
_____
_____
_____
_____
_____
_____
_____

Approved_____Date_____

Photos Attached? Yes ___ No ___

<u>Securing the Outdoor Construction Site—Kevin Wright Carney—Ref: Chapter 4.</u>

(Page 2 of 2)

EMPLOYEE HIRING WORKSHEET

As mentioned in Chapters 4 and 9, every organization has to be careful who it "lets in their house" as Pamela Graham would say. This worksheet or any variation of it should help you screen your potential employees. If your company has a separate personnel (human resources) department, then it should already have some similar format to use. If not, here is a simple form to assist you in your hiring process. There should be an employee who is specifically assigned to the administrative end of this process. If not, your efforts most likely will fall by the wayside.

Keep in mind that if an employee you hired commits a crime during the course of his or her duties and you could have known or should have known that this person had a checkered past, you might be liable in part for their actions. Added to that, if that new hire steals from you, you have no one to blame but yourself if you did not do a properly screening.

 # Employee Hiring Worksheet

Applicant Name_____

Application Date_____ Interview Date_____

Date of Birth_____ Place of Birth_____

Male___ Female___ Height_____ Weight_____ Hair_____ Eyes _____

Authorized to Work in Country? Yes ___ No ___ Work Permit Number_____

Position Applied For_____

Application Complete? Yes ___ No ___

If not, explain: _____

Work Experience Verified? Yes ___ No ___ Verification by_____

Gaps in Employment? Yes ___ No ___

Explain_____

Required Licenses? Yes ___ No ___ License Number(s) _____

Applicant Name_____

Criminal History Clear? Yes ___ No ___ Verification by_____

If not clear, is conviction a disqualifying event? Yes ___ No ___

Explain_____

Driver History Clear? Yes ___ No ___ Verification by_____

If not clear, are there any disqualifying violations? Yes ___ No ___

Explain_____

Insurance (If using personal vehicle) Yes ___ No ___ Verification by_____

References Checked? Yes ___ No ___ Verification by_____

Negative Reports from References? Yes ___ No ___

Explain_____
_____

Securing the Outdoor Construction Site—Kevin Wright Carney—Ref: Chapters 4 and 9.

(Page 1 of 2)

 **Employee Hiring Worksheet**

Applicant Name_____

Credit History Checked? Yes ___ No ___ Verification by_____
Credit History Clear? Yes ___ No ___
If not clear, are there any disqualifying entries? Yes ___ No ___
Explain_____
Education/Training Verified? Yes ___ No ___ Verification by_____
Is all education and training as stated on application? Yes ___ No ___
Explain_____
Other Relevant
Information_____

Approved for Hire? Yes ___ No ___
Hire Date _____ Training and Orientation Date_____
Approved by_____Title_____

Securing the Outdoor Construction Site—Kevin Wright Carney—Ref: Chapters 4 and 9.

(Page 2 of 2)

INJURY REPORT

It is imperative to document any injuries on a work site. This documentation will serve to prove your due diligence if the injury becomes a matter of a claim or civil litigation. Furthermore, if you live in a highly regulated jurisdiction, such as California, there could even be civil and criminal penalties for failure to report serious injuries to the government.

Your company operating manual should have clearly written guidelines for action at the time a person is injured or becomes suddenly ill on site. It is more than prudent to have personnel on site at all times who are trained in first aid and cardiopulmonary resuscitation (CPR). This goes beyond any legal requirement but rather speaks to the moral imperative of caring for injured or suddenly ill persons. Even the best paramedics in the world cannot do any good if the patient is dead when he or she arrives.

Both your insurance provider and your attorney will want to have a record of what efforts were made to save a victim's life or ease suffering during such an incident. Without documentation, the temporary care you avail the patient could be disputed later, and you would have no way to counter allegations of misconduct or indifference.

It is important to get as much information as possible immediately after first aid is rendered. The time to get information is when it is fresh in people's minds. Don't forget to identify witnesses and obtain statements. This might protect you if a witness changes his or her story later. As always, photographs are highly recommended to be included with these reports.

Injury Report

Date of Injury_____ Time_____ Report Number_____

Location Address_____

Patient's Name_____

Patient's Address_____

Patient's Telephone Nr._____ Email Address_____

Employee? Yes ___ No ___ If yes, Employee Number_____ Hire

Date_____

Fatality? Yes? ___ No ___ Rescue Response? Yes ___ No ___ Fire

Department? Yes ___ No ___

Fire Department Engine Company_____ Captain_____

Time:_____

Coroner Response? Yes ___ No ___ Deputy Coroner_____

Time:_____

Law Enforcement response? Yes ___ No ___ Crime? Yes ___ No ___

Law Enforcement Officer_____ ID Number_____

Report Number_____

Nature of Incident Resulting in Injury_____

Injuries Sustained_____

Patient's Name_____ Report Number_____

Multiple Patients? Yes ___ No ___ (Note: Each patient requires a separate
report)

First Aid and/or CPR rendered? Yes ___ No ___ By_____

Patient Transported to Hospital? Yes ___ No ___ Time? _____

Hospital _____

Ambulance Company_____ Phone Nr._____

Narrative_____

_____

_____

Report by _____ Employee Nr. _____

Date_____ Time_____ Approved by_____

Securing the Outdoor Construction Site—Kevin Wright Carney—Ref: Chapter 4.

SITE SURVEY: VULNERABILITY ASSESSMENT REPORT

This report documents the results of your site visits and meetings with construction, law enforcement, and fire professionals. It should tell the client of the realities of security challenges at the site and detail your strategies for providing sound protection for the project.

Although not required, I suggest studying the *ASIS Protection of Assets* manual and perhaps obtaining your Certified Protection Professional (CPP) or Physical Security Professional (PSP) certification. This will assist you in developing the mindset for making these assessments.

As I mentioned in Chapter 5, make sure to actually visit the site and walk the whole area if physically possible. It will help to have a copy of the site plans with you so you can envision how the site will appear to the personnel who are securing it. Study the traffic patterns and likely access of criminals who may approach the site. Think about what barriers and signage might need to be in place to create the impression that the site is protected space and ought not to be trespassed upon.

Think beyond security concerns to physical security and look at safety hazards as well. Think about such natural phenomena as landslides, mudslides, flooding, and forest fires. If such hazards exist, identify escape routes so that all personnel may be safely evacuated in times of danger.

This report will help you, or whomever is tasked to do so, to write an accurate and effective security plan for the site.

Site Survey: Vulnerability Assessment Report

Project_____Job Number_____

Construction Company_____

Location_____

Project Manager_____

Telephone_____ Email Address_____

Fax_____

Security Manager_____

Telephone_____ Email Address_____

Fax_____

Security Company_____

Start Date_____ Estimated Date of

Completion_____

Law Enforcement Agency_____

Telephone_____ Email Address_____

Fax_____

Fire Department Agency_____

Telephone_____ Email Address_____

Fax_____

By _____ Telephone_____

Email_____

Introduction_____

_____

_____

_____

_____

_____

_____

_____

_____

Conference with Site Manager? Yes ___ No ___

Brief Substance of Conference

_____

_____

_____

_____

_____

Plans Reviewed? Yes ___ No ___ Site Walk? Yes ___ No ___ Photographs Taken?
Yes ___ No ___

Law Enforcement Conference? Yes ___ No ___ Fire Department Conference?
Yes ___ No ___

Securing the Outdoor Construction Site—Kevin Wright Carney—Ref: Chapter 5.

(Page 1 of 3)

Site Survey: Vulnerability Assessment Report

Law Enforcement Conference

Officer Interviewed_____

Agency_____ Telephone_____

Email _____

Brief Substance of Conference

_____

_____

_____

_____

Fire Department Conference

Fire Officer Interviewed _____

Department_____ Telephone_____

Email _____

Brief Substance of Conference

_____

_____

_____

_____

Site Walk

Traffic: Heavy___ Moderate___ Light___ Vehicle Speeds_____

    Concrete Barriers Needed? Yes ___ No ___ Signage Needed? Yes ___ No ___

Natural Hazards? Yes ___ No ___ (Check all that apply.)

    Fire___ Landslide___ Mudslides___ Earthquake___ Storms___ Flooding___

    Lightning___ High Winds___ Wild Animals___ Other ___

    Structures on site to be demolished? Yes ___ No ___ Vacant? Yes ___ No ___

    Trees to be felled? Yes ___ No ___ Other Hazards to be removed? Yes ___ No ___

Site Walk Narrative

_____

_____

_____

_____

_____

_____

_____

_____

_____

_____

_____

Securing the Outdoor Construction Site—Kevin Wright Carney—Ref: Chapter 5.

(Page 2 of 3)

Site Survey: Vulnerability Assessment Report

Equipment Needed

Security Fencing? Yes ___ No ___ Lighting? Yes ___ No ___ Camera System? Yes ___ No ___

Alarms? Yes ___ No ___ Signage? Yes ___ No ___

Patrol Vehicle? Yes ___ No ___ Car___ 2WD Truck___ 4WD Truck___ N/A___

Security Personnel Needed

Number of Security Officers per Shift_____ Site Supervisor/Post Commander Yes ___ No ___

Number of Shifts ____ Day___ PM___ EM___

Armed ___ Unarmed___ Firearm___ Tear Gas___ Pepper Spray___ Baton___

Other Considerations

Surrounding Population Urban___ Suburban___ Rural___ Uninhabited___

Theft Vulnerability High___ Strong___ Moderate___ Slight___ Remote___

Fire Danger High___ Strong___ Moderate___ Slight___ Remote___

Other Crime Vulnerability High___ Strong___ Moderate___ Slight___ Remote___

Distance to Police Department_____ Distance to Fire Department_____

Conclusions

_____
_____
_____
_____
_____
_____
_____
_____
_____
_____
_____
_____

Recommendations

_____
_____
_____
_____
_____
_____
_____

Securing the Outdoor Construction Site—Kevin Wright Carney—Ref: Chapter 5.

(Page 3 of 3)

AFTER-ACTION REPORT

This report is primarily for use by both construction and security professionals, although it can be reviewed by law enforcement, insurance, and legal professionals. Its primary purposes are to evaluate the general success of the security plan for each site and assist in developing better plans for future projects. It should be drafted by the security provider and reviewed by the construction site manager. This will serve to keep the report as candid and honest as possible. The report should detail both successes and failures, as well as any unusual occurrences.

I recommend keeping these reports all on file with the original security plan so that when you have a similar project, you won't have to reinvent the same program. There will, of course, be variations on every plan, but having a file copy will point you in the right direction for the next project.

After-Action Report

Project_____Job Number_____
Construction Company_____
Location_____
Project Manager_____
Telephone_____ Email Address_____
Fax_____
Security Manager_____
Telephone_____ Email Address_____
Fax_____
Security Company_____
Start Date_____ Date of Completion_____
Security Planner_____
Telephone_____ Email Address_____
 Fax_____
Description of Project_____
Reports Attached: Site Survey/Vulnerability Assessment___ Security Plan___
Materials Checklist___
Serious Incident Report(s)___ Injury Reports___

Planned Security Budget_____ Actual Expenditure_____
Cost Savings _____ Cost Overruns_____ N/A___
Number of Accidents _____ Initial Cost _____ Less Insurance _____
Total Cost_____
Number of Injuries _____ Initial Cost _____ Less Insurance _____
Total Cost_____
Notable Security Successes:

_____
_____
_____
_____
_____

Notable Security Failures:

_____
_____
_____
_____
_____

Near Misses:

_____
_____
_____

Securing the Outdoor Construction Site—Kevin Wright Carney—Ref: Chapter 5.

After-Action Report

Report Narrative:

_____
_____
_____
_____
_____
_____
_____
_____
_____
_____
_____
_____
_____
_____
_____
_____
_____
_____
_____
_____
_____
_____
_____
_____
_____
_____
_____
_____

Conclusions_____
_____
_____
_____
_____
_____
_____
_____
_____
_____
_____
_____
_____
_____

**Recommendations for Future
Projects**_____
_____
_____

By_____ Date_____

Reviewed By_____ Date_____

Securing the Outdoor Construction Site—Kevin Wright Carney—Ref: Chapter 5.

(Page 2 of 2)

LOSS MITIGATION ACTION PLAN

As mentioned in the book, bad things can and will happen during the course of a construction project. Some will be minor, and some will be potentially catastrophic in nature. The difference between recovery and collapse often lies in the preplanning for a loss event.

Thinking these things out in advance, setting up fast equipment and materials replacement, and developing business alliances will go a long way toward mitigating the impact of a major event. Certainly, insurance is an integral part of any mitigation plan, but it should not be the total plan. In the case of a major catastrophe wherein a lot of claimants are involved, it may take some time before your insurance provider can adequately respond to your claim, hence the need for a cash reserve and a sound plan to get your project back in operation as soon as possible.

Remember that if an incident is a severe natural disaster, your employees will most likely have personal business to settle before reporting to work. If you have set aside money in your cash reserve to assist them until their own insurance can help them, they can get their families settled for the time being and be ready to return to work. You may need to arrange such simple tasks as carpools for those who might have lost the use of their personal vehicles.

It is wise to contact your insurance agent to ascertain if your mitigation plan can be kept on file for sharing with a claims adjustor when a loss event occurs.

 **Loss Mitigation Action Plan**

Project _____Job Number_____

Construction Company_____

Location_____

Project Manager_____

Telephone_____ Email Address_____

Fax_____

Security Manager_____

Telephone_____ Email Address_____

Fax_____

Security Company_____

Start Date_____ Scheduled Date of

Completion_____

Insurance Company_____ Telephone _____

Insurance Agent_____ Policy Number_____

Nature of Loss_____

Personnel Plan:

Personnel Roster Attached? Yes ___ No ___

Critical Positions_____

Personnel Available from Other Sites

_____

_____

_____

_____

_____

Employment Office_____Telephone_____

Union Hall_____Telephone_____

Alliance Partner_____Telephone_____

Other Source(s)_____Telephone_____

Employee Assistance Fund Amount $_____

Assigned Fund Administrator_____

Allowable Expenditures: Medical Emergencies___   Hotel___ Vehicle Repair___

Tools___

Emergency Home Repair___ Transportation___ Food___ Clothing___

Other (Specify)_____

Level of Approval Required: Owner___ General Manager___ Site Manager___

Fund Administrator___

Personnel Transportation: Personal Vehicles___ Company Van___ Bus___

Taxi___ Other___

Securing the Outdoor Construction Site—Kevin Wright Carney—Ref: Chapter 6.

(Page 1 of 3)

 # Loss Mitigation Action Plan

Mutual Aid Alliances:

Company_____ Telephone_____
Equipment___ Personnel___ Materials___ Contact Name_____
Company_____ Telephone_____
Equipment___ Personnel___ Materials___ Contact Name_____
Company_____ Telephone_____
Equipment___ Personnel___ Materials___ Contact Name_____
Company_____ Telephone_____
Equipment___ Personnel___ Materials___ Contact Name_____
Company_____ Telephone_____
Equipment___ Personnel___ Materials___ Contact Name_____
Company_____ Telephone_____
Equipment___ Personnel___ Materials___ Contact Name_____

Monetary Reserve: $_____

Assigned Fund Administrator_____

Allowable Expenditures: Additional Personnel ___ Heavy Machinery Rental___
Equipment Rental___ Transportation___ Vehicle Rental___ Mutual Aid Response
to Alliance Partners___ Materials___

Meals___ Lodging___ Additional Security___ Security Equipment___

Other (Specify) _____

Level of Approval Required: Owner___ General Manager___ Site Manager___
Fund Administrator___
Notifications:

Insurance Agent_____ Telephone_____

Time Notified_____

Attorney_____ Telephone_____

Time Notified_____

Alliance Partner_____ Telephone_____

Time Notified_____

Securing the Outdoor Construction Site—Kevin Wright Carney—Ref: Chapter 6.

(Page 2 of 3)

 # Loss Mitigation Action Plan

Equipment Replacement:

Can equipment on site be used for alternate tasks until correct equipment arrives?
Yes ___ No ___

Can equipment from another company site be used? Yes ___ No ___

Rental Company_____Telephone_____

Prearrangements Made? Yes ___ No ___   Contact_____

Rental Company_____Telephone_____

Prearrangements Made? Yes ___ No ___
Contact_____

Rental Company_____Telephone_____

Prearrangements Made? Yes ___ No ___   Contact_____

Materials Replacement:

Are missing materials readily available from standard sources? Yes ___ No ___

Can materials from another company site be used? Yes ___ No ___

Materials Supplier_____Telephone_____

Prearrangements Made? Yes ___ No ___   Contact_____

Materials Supplier_____Telephone_____

Prearrangements Made? Yes ___ No ___   Contact_____

Materials Supplier_____Telephone_____

Prearrangements Made? Yes ___ No ___   Contact_____

Materials Supplier_____Telephone_____

Prearrangements Made? Yes ___ No ___   Contact_____

Media/Public Relations Spokesperson_____

Title_____ Telephone_____

Level of Approval Required for Info Release: Owner___ General Manager___ Site
Manager___

Securing the Outdoor Construction Site—Kevin Wright Carney—Ref: Chapter 6.

(Page 3 of 3)

CONTRACT SECURITY VERSUS PROPRIETARY SECURITY WORKSHEET

Some construction professionals always use contract security companies to protect their projects, and some always use their own personnel for security (i.e., proprietary security). This worksheet is for the construction professionals who are undecided and wish to weigh the two concepts.

Of course, coming from the security field, I suggest using a professional security company. Security is the reason they exist, and they are not distracted by the pressing concerns of building a large outdoor capital project.

If you are compelled to use your own personnel, then I think it is wise to use a separate security-dedicated staff to provide protection, to the exclusion of all other crew members. This means you don't use laborers or equipment operators to provide security (other than for slow-speed traffic control). Included in that equation is the requirement that your security staff either come from the security industry or that you send them to a security officer school for proper training.

When using the worksheet, keep in mind that most times the costs enumerated on the form are included in the contract price of the security company contract. When you see how much is included in the contract, you will most likely see a cost savings in contracting for security service.

Contract Security vs Proprietary Security Worksheet

Consideration	Contract Security	Proprietary Security
Security Training?	Yes ___ No ___	Yes ___ No ___
Security Licensed?	Yes ___ No ___	Yes ___ No ___
Security Insured?	Yes ___ No ___	Yes ___ No ___
Security Supervised?	Yes ___ No ___	Yes ___ No ___
Security Equipped?	Yes ___ No ___	Yes ___ No ___
Security Vehicles?	Yes ___ No ___	Yes ___ No ___
Officers Fatigued?	Yes ___ No ___	Yes ___ No ___
Officer Hourly Cost	$_____	$_____
Supervisor Hourly Cost	$_____	$_____
Vehicle Hourly Cost	$_____	$_____
Equipment Hourly Cost	$_____	$_____
Hourly Costs × Nr of Hours	$_____	$_____
Best Value (Check One)	___	___

Securing the Outdoor Construction Site—Kevin Wright Carney—Ref: Chapter 7.

PROPOSED SECURITY CONTRACTOR WORKSHEET

If you have a years-long relationship with your security provider and are happy with them and their pricing and service, then good for you. Long-term business partnerships are valuable.

Perhaps, however, you are contracting for security service for the first time or you are dissatisfied with your current security provider. In that case, this worksheet will help you select a company to protect your jobsite.

Remember that even with the best of security programs, occasionally a thief will be successful against your company's assets. However, you can greatly reduce that possibility by selecting the most professional company from the start and giving them the latitude to provide you with their best security program.

You should complete a form for each of the companies that you are considering for your security contract. Then compare the forms side by side. Keep in mind that price should not be your only consideration, although it is important. Sometimes if the bids are not that far apart, you can negotiate the price with the company that you like best. Remember the old adage "you get what you pay for." The extra pennies that you spend for an excellent security provider can save you tens of thousands of dollars, pounds, or Euros in the long run.

This worksheet has a point-based scoring system that will help you score your security bidders. The choice is up to you, so take time and care when deciding. Some things such as lack of insurance can be a disqualifier, regardless of how they score on the rest of the worksheet.

Proposed Security Contractor Worksheet

Project _____Job Number_____
Construction Company_____
Location_____
Project Manager_____
Security Proposal Contact _____ Position _____
Telephone _____ Email Address _____
Fax _____
Security Company _____
Start Date _____ Scheduled Date of Completion _____

1. Service approach: Can they meet our needs? Yes ___ No ___

2. Do they present a clear picture of how their services are managed/supervised? Yes ___ No ___

3. Is their management easy to contact when needed? Yes ___ No ___

4. Is their hiring locally based? Yes ___ No ___

5. Have they explained their supervision program? Yes ___ No ___

6. Have they explained their report writing procedures? Yes ___ No ___

7. Is their notification to you in times of emergencies satisfactory? Yes ___ No ___

8. Have they explained how their security officers are screened and hired? Yes ___ No ___

9. Have they explained how their security officers are trained? Yes ___ No ___

10. Have they explained their experience requirements? Yes ___ No ___

11. Have they discussed how security officers' activities are tracked? Yes ___ No ___

12. Do they have adequate coverage for vacations and sick call-ins? Yes ___ No ___

13. Do they have additional personnel if needed? Yes ___ No ___

14. Do they have adequate insurance? Yes ___ No ___ *

15. Have the explained their billing process? Yes ___ No ___

16. Have they asked to look at the construction site plan? Yes ___ No ___

17. Have they walked the site? Yes ___ No ___

18. Have they discussed security equipment? Yes ___ No ___

19. Have they discussed lighting? Yes ___ No ___

20. Have they discussed the camera system? Yes ___ No ___

21. Have they discussed patrol checkpoints? Yes ___ No ___

22. Have they discussed communication equipment? Yes ___ No ___

23. Have they discussed secure fencing? Yes ___ No ___

24. Have they presented reasonable pricing? Yes ___ No ___

25. Do you trust them? Yes ___ No ___ *

*A "no" answer might be a disqualifier regardless of the score.

Securing the Outdoor Construction Site—Kevin Wright Carney—Ref: Chapter 7.

(Page 1 of 2)

Proposed Security Contractor Worksheet

Other Observations and Considerations:

_____
_____
_____
_____
_____
_____
_____
_____

Number of "Yes" Answers: _____

Number of "No" Answers: _____

**Total of "Yes" answers minus total of "No" answers × 4 = _____ Score
(100 maximum points)**

Acceptable? Yes ___ No ___

Evaluation by_____ Position_____

Approved by _____

Position _____

Securing the Outdoor Construction Site—Kevin Wright Carney—Ref: Chapter 7.

(Page 2 of 2)

CHAPTER POINTS BY DISCIPLINE

1. **Security professionals:** This chapter has checklists and worksheets that will be used by you, the security contractor, or both. Regardless of who the forms are for, you should be familiar with all of them so you might make the best presentation and give the best security service possible to your client. Also, make your clients aware of these checklists and this book so that your efforts and theirs will be well coordinated.

2. **Construction professionals:** The forms in this chapter are all either designed to be used by you or to be completed by your security provider and given to you. By familiarizing yourself with all of the checklists and worksheets, you will gain a better understanding of what to expect from your security provider. These forms help to bring into focus many of the usable aspects of the concepts in this book.

3. **Law enforcement professionals:** The fact that you have purchased and read this book is an indicator of your commitment to your duties as a law enforcement to the protection of these critical infrastructure construction projects and to the people who dedicate their lives to the construction profession. By understanding the information that is captured on these worksheets, you will have an increased knowledge of the procedures and practices that will make your job easier.

   Furthermore, you will be able to advise and assist those friends and acquaintances in the construction and security professions if they look to you for advice.

4. **Insurance professionals:** The checklists and worksheets in this chapter and this book can help you to advise your construction clients, whether they are construction professionals or security professionals. By helping both of them provide more effective security, you will enhance safety, help elevate construction security standards, and thus enhance the bottom line of your organization.

   You can also assist your clients and the insurance industry by advocating and promulgating the proposed standards set forth in the book. The goal of this book is to reduce construction site crime in a meaningful way, and the insurance industry has been the primary driving force behind this effort.

5. **Legal professionals:** As lawyers, you should understand the concepts and procedures for which these checklists and worksheets are designed to be used. These documents can help your clients document the steps they have taken to prevent crime and mitigate loss. With these, you can encourage your construction clients to stop "rolling the dice" by leaving construction site security to luck.

   Furthermore, by encouraging the use of these forms and the principles in this book, you can assist your security provider clients by giving them documentable evidence that they have exercised proper care and an abundance of caution when preparing for the security of outdoor capital construction projects.

The National Equipment Register and National Plant and Equipment Register

13

This chapter discusses the National Equipment Register in the United States and the National Plant and Equipment Register in the United Kingdom and Europe. It describes the enormous benefit that these organizations have to the construction industry, law enforcement, and the insurance industry. Security and legal professionals should be aware of these organizations in order to help advise clients on the importance of heavy equipment registration.

THE NATIONAL EQUIPMENT REGISTER (UNITED STATES)

As I said earlier in the book, when the government does not respond, often the private sector will act in order to fill a void. I believe that is a good thing. Such is the case in the United States with the National Equipment Register (NER) and equally so with National Plant and Equipment Register (TER) in the United Kingdom and Europe and the NER in Australia. Whereas this chapter is not an advertisement for the NER, you will see by the time that you reach the end that this organization is valuable to the construction and insurance industries, as well as to law enforcement.

Primarily driven by forces in the insurance industry, the NER was formed as a subsidiary of its parent company, Verisk Crime Analytics. I spoke at length with Ryan Shepherd, the general manager of the NER, to gain greater insight into the company's goals and services. He explained the services of the NER and how it represents a value to its clients. The thing that I noted at the end of our interview was how inexpensive the NER's services are. However, when you balance these minimal costs against the potentially disastrous results of the theft of even one piece of heavy equipment, they are negligible.

Clients can, for a very nominal fee, register their construction equipment with NER just as one would register a car or truck with the local jurisdiction's department of motor vehicles. If the equipment has no Vehicle Identification Number, the NER will assign one and advise the client where to etch or stamp the number onto the machine. The registration information is then available to law enforcement to determine ownership. Theft reports, serious damage reports, and fire reports are also added to the equipment's record.

Also with the registration packet, NER issues antitheft decals to warn thieves that the equipment is registered and can be tracked. If the thieves are professionals, these decals might lead them to select another piece of equipment to steal.

IRONwatch is an inexpensive vehicle tracking system that not only assists in locating heavy equipment that is misplaced on the construction site but also assists law enforcement in locating the stolen machinery. The unique feature of this system is that you can locate machinery without necessarily notifying law enforcement. Why is that important? Sometimes equipment is just mislaid and not stolen.

Using this system, one can "ping" the equipment to locate it in an instant. Also, if a newly hired heavy equipment operator arrives on site without knowing where his or her equipment is, he or she can ping the equipment and save the contractor expensive wage monies while searching for the machine. Also, the system provides three daily 'heartbeats" that provide a snapshot of the vehicle's location.

This system provides location only, but for a slight extra charge, additional telematics can be added to track fuel costs and other expenses associated with operating the machinery.

This service is valuable to rental companies as well as construction contractors to help prevent rented equipment from being lost or stolen. It also helps locate equipment that has not been returned at the end of the rental contract. The equipment can be tracked to 60 countries worldwide.

IRONcheck is the system that most interested me. This is the closest thing in the United States to a stolen equipment database for construction heavy equipment. Through this system, law enforcement, insurers, equipment dealers, and prospective equipment buyers can see if a piece of equipment has been reported stolen. Also, when one of the NER-registered pieces is put up for sale, the transaction is verified with the equipment's owner to make sure that the equipment is in fact, for sale. Ryan Shepherd explained that a current trend among thieves is to sell the equipment before the theft occurs. Then both the client and law enforcement get a "head's up" that a theft is potentially imminent.

Furthermore, the data in IRONcheck can assist insurance adjustors to figure out a fair price for a piece of equipment to settle a claim. The insurance adjuster can check to see if the equipment has had multiple owners and to see what kinds of companies owned the equipment.

This information is also valuable to potential buyers of a piece of machinery because by looking at the ownership records, a potential buyer can discern the probable wear and tear that the machine might have had and then make a fair offer to the seller based on that information. At times, used heavy equipment dealers and auction houses let buyers know of NER's services as a courtesy.

NER also offers two ancillary risk solutions programs for its clients. One is proprietary, and the other is not. **Intellicorp** is an employee background screening service that is a partner company, also owned by Verisk Crime Analytics. If a company has no hiring screening program, this program is as good as any other company out there, and Intellicorp offers discount pricing to NER members.

Another business ally of NER is Escrow.com. This company has a number of programs, but of most interest to construction equipment buyers is the feature that allows buyers to put purchase funds in escrow, much like when buying a house. When the vehicle arrives, and the buyer agrees that it is as was advertised, then the buyer may release the funds from escrow to the seller. This service is particularly significant when purchasing equipment from out of state or from another country.

Again, this chapter is not a business endorsement for Intellicorp or Escrow.com but rather demonstrates NER's commitment to assisting the industry.

NER has a unique public–private relationship with law enforcement. Any bona fide law enforcement officer in the country can set up his or her own account free with the NER. With this membership, they can access NER's registration information, even from a laptop computer in a radio car. That way, if they see a piece of equipment being moved in the middle of the night or notice a piece of construction equipment that has been sitting in one place for an extended period of time, they can check the data base themselves (or through their department's dispatch) to see if it is stolen. This check can be made either via the website or by telephone.

This can lead to either a great observation arrest or the recovery of a stolen piece of construction machinery. Either way, it's a feather in the cap for the officer and his or her agency and a welcome recovery for the construction company or the insurance company, which has already paid a claim.

NER cooperates with law enforcement officials in Canada and Mexico and assists them with their investigations or with theft follow up for machinery that is stolen in the United States and brought to those countries.

Furthermore, NER presents "Heavy Equipment Summits" wherein law enforcement, insurance, and construction professionals are invited to attend. These conferences cover a variety of topics and discuss "red flag scenarios" to help attendees identify circumstances that are precursors to a theft.

NER has a relationship with more than 25 major insurance companies in the United States and a myriad of its subsidiary companies. The insurance industry has a major role in driving the efforts of the NER. In addition to providing support and impetus to NER's efforts, many of those insurance companies offer machinery theft deductible waivers to clients who register and track their machinery through NER. Some of these waivers are as much as $10,000.

Construction equipment lenders also avail themselves of NER's services, especially when flooring equipment at dealerships. They are aware that thefts can occur at the dealership, as well as the construction site.

Ryan Shepherd left me with the thought that every construction company should have a security plan as a part of its overall business structure. He agreed with me that construction professionals should stop "rolling the dice." He said, "Put your money where your mouth is and follow through. You will see results if you stay the course." I agree wholeheartedly.

THE NATIONAL PLANT AND EQUIPMENT REGISTER (UNITED KINGDOM AND EUROPE)

TER provides similar services as NER for the United Kingdom and Europe. TER's admonition is "You cannot buy good title to stolen property, no how many hands it has been through." This is true, pretty much throughout the world. If you buy something that has been stolen, you will have to give it back to the rightful owner, no matter how much you paid for it.

Buying stolen equipment can ruin a company's reputation even if the buyer believed the transaction was legitimate. TER's programs help avoid that pitfall and assists equipment owners, law enforcement, and insurance companies in recovering stolen heavy equipment.

TER's system does not contain the satellite tracking feature that NER has, but that should not be a deterrent for membership. Tracking systems can be obtained from other vendors. The key to TER's great value is the vehicle registration system and some unique partnerships that is has with law enforcement.

Currently, TER has more than 1 million pieces of equipment registered, with a combined value of approximately $7,000,000,000.

One of the great services of TER is that within 24 hours of thefts that exceed $30,000 in value, TER circulates the details of the theft to law enforcement agencies and port authorities throughout its service area. The TER shares information and works together with the Motor Insurer's Anti-Fraud register (MIAFTR).

Heavy equipment can be registered with TER by equipment owners and users, law enforcement, insurance brokers, and insurance company claims departments, as well as individual claims adjusters. This data are then made available to law enforcement 24/7.

TER also issues a TER check certificate upon request for buyers, equipment dealers, and auction houses. For the sellers, this certificate can show that you exercised "due diligence" to establish ownership verification of equipment and machinery.

TER collects its own theft data, which are shared with the police. They employ part-time contract "specialist plant investigators" who initiate theft and fraud investigations based on information and trends gleaned for the NER data and intelligence. TER conducts its investigations and turns the results over to law enforcement for further investigation and prosecution.

TER offers as a service to law enforcement the "Equipment Theft Awareness" program, which consists of three parts: briefings, onsite trainings, and materials.

The briefings cover the nature and extent of plant and equipment theft, the services provided by TER for the police, and targeted information about current threats.

The onsite trainings are conducted in plant yards, at roadside checks, and in operations. The idea is to show officers a range of equipment in the working environment and how to identify them. It appears that this concept gives police officers an accurate feel for how construction sites operate and helps them to identify the vulnerabilities for the equipment that is being operated and stored there.

The materials include Plant & Equipment Identification Guides that identify equipment identification number locations, as well as targeting and threat information. They also include posters, contact cards, newsletters, and the annual TER *Equipment Theft Report*.

An additional service that TER provides to the insurance industry is a selling service for recovered stolen machinery for which claims have been paid. These sales are bid sales, in which the equipment is sold to the highest bidder.

THE NATIONAL EQUIPMENT REGISTER (AUSTRALIA)

NER Australia was formed by two organizations, the DataDot Technology Ltd. and Crimestoppers. Its function is similar to the U.S. NER and the TER in Britain. It provides for equipment registry and cooperate with the Australian Federal Police as well as state and local agencies.

The unique feature of this organization is the tracking system, which uses a commercial product called "DataDot DNA." These DataDots are programmable electronic fingerprint devices that are no larger than a grain of sand. They are either hand painted or sprayed onto various parts, both exposed and hidden, of a piece of machinery. They provide indelible marking to prove to whom the machine belongs.

CHAPTER POINTS BY DISCIPLINE

1. **Security professionals:** As security professionals, part of your duties include keeping your clients apprised of systems that will protect them and their equipment. For that reason alone, you should be aware of the equipment registers in your local jurisdictions. Sharing this knowledge with your clients will enhance their perception of you as a valuable business partner rather than just someone who is hired to keep the boogeyman away. If you are assigned by your client to act as an agent between the client and the NER, you should know how the registration and tracking systems work. They will make your job infinitely easier.
2. **Construction professionals:** The concept of the NER should be as familiar to you as any of the equipment and techniques that you use to build your projects. Registering your equipment into a database that law enforcement and insurance companies use will vastly enhance the probability that equipment stolen from your sites will be recovered and returned.

    Furthermore, because most construction equipment does not come with a title and there are no government registers, recording your ownership of equipment with the NER might be your only proof that your equipment belongs to you. Added to that, the liaisons among the NER, law enforcement, and your insurance company make membership indispensable.

If you live in a country where no NER exists, it would behoove you to partner with the insurance companies, equipment dealer associations, law enforcement professionals, and other construction professionals to establish such a register where you live.

3. **Law enforcement professionals:** No matter the size of your law enforcement agency, large or small, you can and should join the NER's roster of professionals. Membership is free for law enforcement professionals, and you can access a wealth of data on heavy construction equipment, theft status, and registration and ownership status.

NER and TER both put on training for law enforcement officers and provide free information on their websites to assist in your efforts to stem crime. During the course of these trainings, you will develop relationships that will allow you to network with law enforcement beyond the limits of your jurisdiction.

The people who run the NER and TER are trustworthy, and you can rely on their information to be as accurate as the information that is reported to them by their clients and law enforcement; their database fills the void of government, where little usable data exist.

4. **Insurance professionals:** Because the equipment register concept comes from the insurance industry, most of you who are involved in insuring heavy equipment, or settling claims for that which has been stolen, are already familiar with the services provided by the NER and TER. However, if you have not yet availed yourself of their services, it would be wise to become a member. You will become aware of many valuable free services that are available to you and make some great contacts.

5. **Legal professionals:** If you are a lawyer, either civil or criminal, and you practice law that revolves around the theft of construction equipment and material, you already know that you are in a very small circle of in-demand legal professionals. Chances are that you too already know of the NER, but if not, you should become familiar with it.

Your clients rely on your detailed and intimate knowledge of every aspect of the construction industry. Knowing about the NER and its functions will not only help you with cases but will also help you provide your clients with wise counsel as to how to protect their heavy equipment.

Appendix: 2013 Theft Report

OUR PURPOSE
AN ALLIANCE WITH A PURPOSE

Through a joint alliance, the National Equipment Register (NER) and the National Insurance Crime Bureau (NICB) continue to make life more difficult for equipment thieves. By combining services and areas of expertise, we're providing an efficient conduit for law enforcement and insurers to identify any type of heavy equipment at any time of day and to help contractors reduce the likelihood of unknowingly purchasing stolen equipment.

Our alliance ensures that NER will continue to provide, manage, and expand its database of insurer-supplied theft reports and information about manufacturers, owners, and damaged equipment. NICB will extend the reach and value of that information through its nationwide network of special agents, who are trained in heavy-equipment theft and available to respond to law enforcement calls for investigative assistance or identification requests.

Better ownership documentation, accurate equipment identification, proper reporting, and greater site security will continue to increase the ability of law enforcement to combat equipment theft. Awareness, education, and training are key components of an overall fraud-prevention plan that may lead to immediate economic benefits for contractors, owners, and insurers.

Through our joint efforts, we're reducing the cost of theft for equipment owners and insurers by increasing the likelihood of recovery and arrest. We're also limiting the ability to fence stolen equipment, thus making heavy equipment a riskier target for thieves.

National Equipment Register 545 Washington Boulevard Jersey City, NJ, 07310-1686 201-469-2030 info@ner.net www.ner.net.

National Insurance Crime Bureau 1111 East Touhy Avenue, Suite 400 Des Plaines, IL 60018 847-544-7000 www.nicb.org.

INTRODUCTION
OVERVIEW
The National Equipment Register (NER) and National Insurance Crime Bureau (NICB) annual report on equipment theft in the United States is based primarily on data the NICB drew from the National Crime Information Center's (NCIC) database of more than 10,000 thefts of construction and farm equipment in 2013 and information reported to ISO ClaimSearch.® We'll publish similar reports every year to help track trends using the growing volume of data available to NER and the NICB.

AIM
Our study provides equipment owners, insurance companies, and law enforcement with information to guide theft-prevention efforts and allocate investigative resources. The study puts the information into context through footnotes, analyses, and conclusions that relate to the protection, investigation, and recovery of heavy equipment.

As in the past, the 2013 report seeks to answer key questions: Who steals heavy equipment, and how do they do it? How much and what types of equipment do they steal? Where do they steal equipment from, and where does it go?

DATA SOURCES
The NICB has access to all the data in the NCIC vehicle theft file, and it maintains a mirror image of that file. The FBI and other federal, state, local, and foreign criminal justice agencies as well as authorized courts submit data on stolen vehicles, stolen vehicle parts, and mobile off-road equipment and components. The NICB uses the data to assist insurance companies in recovering stolen vehicles and mobile off-road equipment.

Since 2001, NER has developed databases of heavy-equipment ownership and theft information. Owners and law enforcement agencies report thefts directly to NER's database through its website. Insurers report thefts through ISO ClaimSearch, the insurance industry's all claims database. Through an alliance with the American Rental Association (ARA), NER can capture loss and ownership data from many of the world's largest rental fleets and hundreds of smaller fleets.

Although statistics can't reveal all underlying reasons for the high level of equipment theft, we can draw conclusions from trends and the daily contact that NER staff members have with theft victims, insurers, and law enforcement.

PRESENTATION AND ANALYSIS

We've presented each set of data in graphs or tables to allow easy comparison and to highlight trends. Notes explain data sources and gathering techniques. Analyses discuss the relative importance of the factors that affect each set of results. We provide additional commentary where results suggest a particular action or response.

THEFT STATISTICS
THEFT BY STATE

Top Ten States for Equipment Theft in 2013

Rank	State	Thefts
1	Texas	1,494
2	North Carolina	913
3	Florida	892
4	California	734
5	South Carolina	691
6	Georgia	609
7	Tennessee	526
8	Oklahoma	525
9	Alabama	398
10	Arkansas	358

Notes
1. Although equipment thefts occurred in every state, the top five states accounted for 41% of the total number of thefts in 2013. In 2012, the top four states accounted for 37%.
2. The table represents 11,486 equipment theft reports captured by NCIC during 2013.

The top five states account for 41% of all thefts.
The top ten states account for 62% of all thefts.

Analysis

1. Theft levels closely correspond to the amount of equipment in a particular area. In other words, the states with the highest volume of construction and agriculture—and therefore the most machinery—have the largest number of thefts.
2. Organized theft rings are likely to develop in areas with a high concentration of equipment and a large number of potential buyers of used equipment, stolen or otherwise. Higher loss ratios for insurers in certain areas reflect that development.

Comment

Sometimes theft hot spots emerge when an organized group of thieves and fences is working in a particular area. NER's regional theft-trend alerts highlight such activity. Detecting and thwarting those groups often coincide with a noticeable drop in theft rates. Documented recoveries illustrate that correlation. Some examples are in the "Case Studies" section.

THEFT BY TYPE OF LOCATION

The graph below shows insured losses by the type of location of the theft:

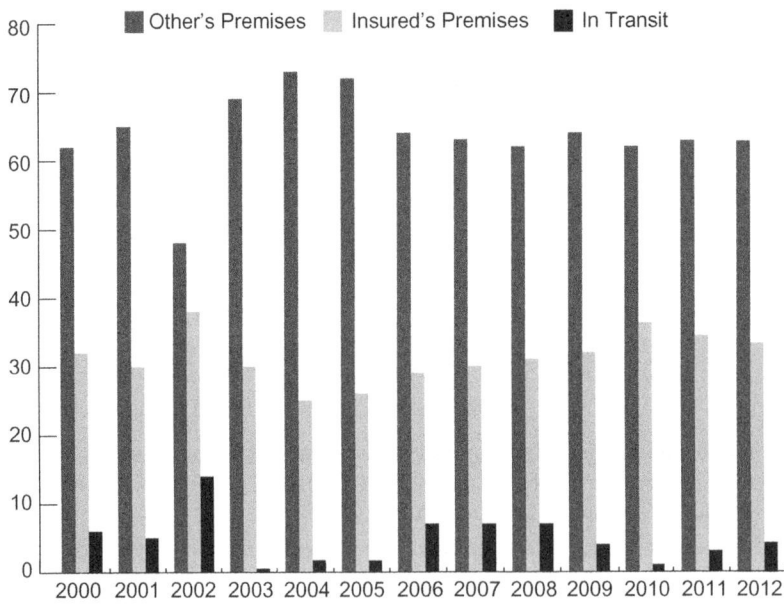

Notes

1. Losses by type of location of theft are displayed as a percentage of all claims.
2. Source is ISO Inland Marine Circular, Contractors Equipment, All Classes.

Analysis

With regard to theft by type of location, two factors should be considered: the location where the equipment spends the most time and the level of security at each type of location. Most often, equipment is on a work site, labeled on the graph as "Other's Premises." Those work sites usually have lower levels of physical security than an "Insured's Premises," which is often a fenced-in compound.

Comment

It's not enough to focus solely on the security of premises and work sites. Equipment users should secure machines, even if they can do so only temporarily. For example, a user could surround mobile equipment with hard-to-move objects when the equipment is not in use.

TYPES OF EQUIPMENT STOLEN

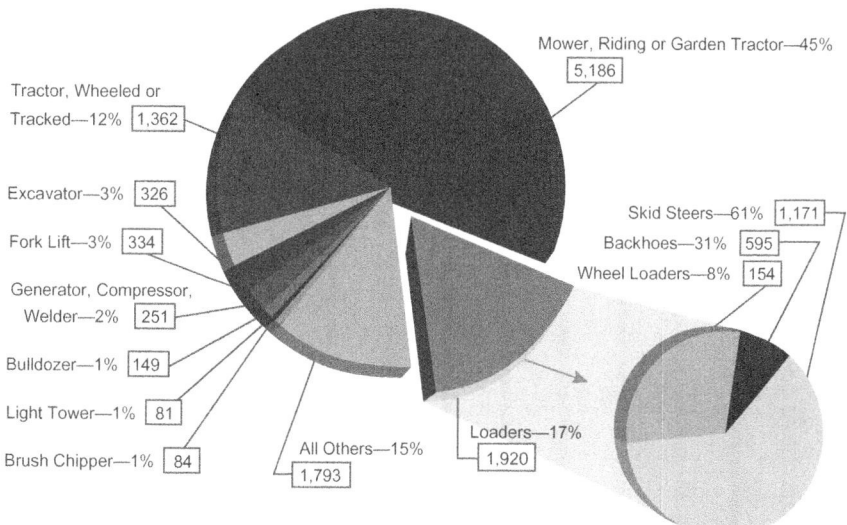

Mower, Riding or Garden Tractor—45% 5,186

Tractor, Wheeled or Tracked—12% 1,362

Excavator—3% 326

Fork Lift—3% 334

Generator, Compressor, Welder—2% 251

Bulldozer—1% 149

Light Tower—1% 81

Brush Chipper—1% 84

All Others—15% 1,793

Loaders—17% 1,920

Skid Steers—61% 1,171

Backhoes—31% 595

Wheel Loaders—8% 154

Notes

1. The chart represents 11,486 theft reports submitted to NCIC in 2013.
2. The inclusion of landscaping equipment—mainly commercial riding mowers—reduces the percentage of all other categories.
3. The top five types of equipment account for 79% of all losses. In 2012, the top five represented 86% of all thefts.
4. "Tractor" is a broad category, including compact, utility, and agricultural tractors.
5. More than 50 types of equipment make up the "All Other" category. They include graders, scrapers, wood chippers, and rollers. Unidentified construction and farm equipment represent the majority (more than 900) of the "All Other" category.

Analysis

1. Two key factors determine the type of equipment that thieves are most likely to steal: value and mobility. Value is the primary factor, except for items too large to move on a small trailer. For instance, large bulldozers are valuable but seldom stolen, as they are difficult to move.
2. Another factor to consider is the number of each type of equipment in circulation. For example, skid steer loaders account for more than 30 percent of new construction equipment sold in the United States in the last five years.
3. Dozers and wheel loaders are the most valuable types of equipment in the top ten, but backhoes and skid steers are easier to transport. Therefore, the latter group represents a greater percentage of thefts.
4. The types of high-value equipment reported stolen frequently are wheeled machines, such as wheel loaders.

Comment

Equipment owners should consider mobility of equipment, as well as value, when planning security efforts.

FREQUENCY OF THEFT COMPARED WITH OTHER RISKS

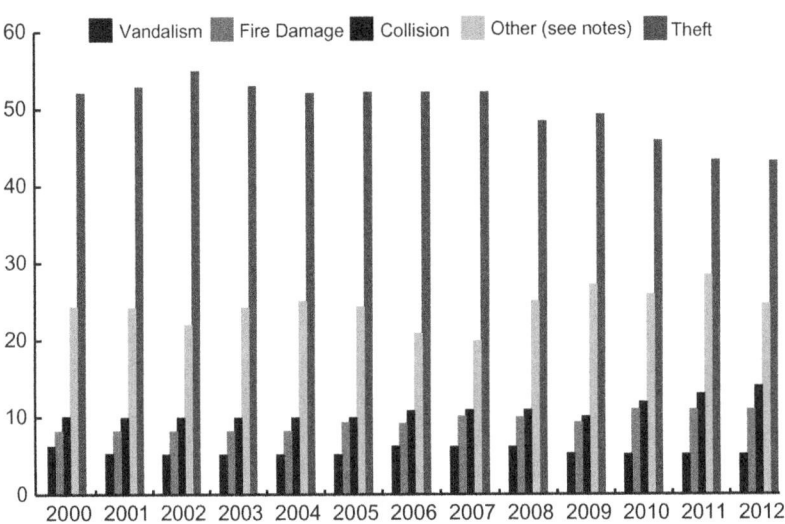

Notes

1. Frequency of risk is displayed as a percentage of all claims.
2. Source is ISO Inland Marine Circular, Contractors Equipment, All Classes.
3. We base the figures on frequency, not value. Theft still tops the list by value, although by a smaller margin.
4. "Other" includes claims involving windstorm, hail, water damage, food, volcanic action, and earthquake.

Comment

Theft is the most frequent cause of loss, but it is also the type of loss that good prevention most dramatically affects. In other words, the level of risk varies greatly between equipment owners who take certain precautions and those who do not. Equipment owners can reduce the likelihood of theft and improve the chances of recovery by taking simple preventive steps that are both cost-effective and measurable.

THEFT BY MANUFACTURER

Manufacturer	Thefts
John Deere	2,445
Kubota Tractor Corp.	1,025
Bobcat	721
Caterpillar	679
Toro	364
Case	308
Husqvarna	306
Craftsman	280
Exmark	275
Cub Cadet Corp.	260

Notes

1. Source is the total number of thefts reported to NCIC during 2013.

Analysis

1. While all makes of off-road equipment have little or no standard equipment security, the manufacturers on the above list make the most compact, and thus most easily stolen, equipment. The list does not necessarily follow the entire market share of all heavy equipment manufactured.
2. If two pieces of equipment are equally easy to steal, a thief is more likely to steal the machine of greater value. Age, condition, and brand determine a machine's perceived value.
3. New results will emerge as manufacturers register sales with NER, work closely with NICB investigators, and include additional security measures as standard features.

THEFT BY MONTH

The graph below illustrates equipment losses by the month the theft was reported.

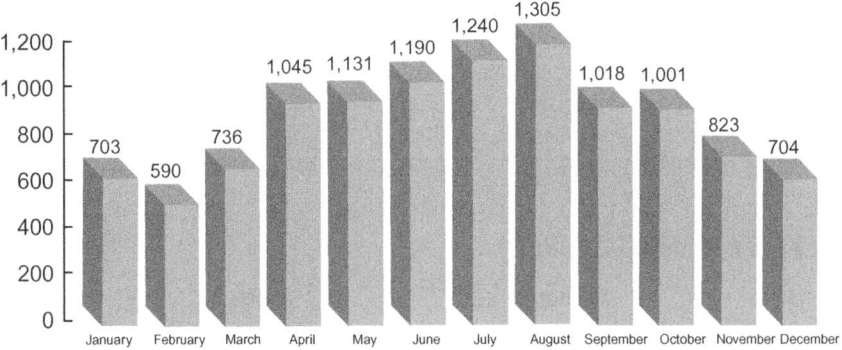

Analysis

Theft levels closely correspond with peak construction periods. In other words, the months with the highest volume of theft are the ones that have increased equipment activity due to cooperative weather, longer days, and the end of a crop growth cycle. As equipment owners move items between jobsites and fields, there are greater risks, exposures, and opportunities for theft. There is an additional likelihood that thefts may go unnoticed for a longer period of time than when equipment is stolen from an owner's yard.

MODEL YEAR OF EQUIPMENT STOLEN

Equipment produced in the last ten years accounted for 74 percent of thefts reported to NCIC in 2013. Forty-eight percent of thefts reported in 2013 were machines manufactured in the last five years. The table lists the top ten years of manufacture for machines stolen in 2013:

Year	Amount
2013	1,806
2012	1,411
2010	880
2011	873
2007	686
2008	636
2005	616
2006	606
2009	505
2004	442

Notes

1. Source is the total number of thefts reported to NCIC during 2013.
2. Each piece of equipment manufactured in 2013 faced potential theft for only part of the year— from the date sold to December 31.

3. Results may be skewed slightly because owners often misstate the date of manufacture. For example, a buyer may list a 2012 model purchased in 2013 as a 2013 model.

Analysis

The newer a piece of equipment, the more likely it is to be stolen. If given the choice between two similar machines, a thief will choose the newer, more valuable machine, because they are equally easy to steal. Those results are in stark contrast to larger trends in automobile theft, where older models account for more stolen cars. Newer cars carry more sophisticated antitheft technology. Heavy-equipment design, however, emphasizes productivity instead of security. The necessity for multiple operators leads to little or no antitheft technology. Many heavy-equipment manufacturers installed as few security features on 2012 models as they did on 1980 models.

TOP 10 CITIES FOR EQUIPMENT THEFT

City	State	Thefts
Houston	TX	199
Oklahoma City	OK	111
San Antonio	TX	82
Miami	FL	77
West Palm Beach	FL	72
Conroe	TX	65
San Diego	CA	65
Charlotte	NC	64
Tacoma	WA	59
Orlando	FL	56

Notes

1. Source is the total number of thefts reported to NCIC during 2013.
2. Nine of the top ten cities are in the top ten states for theft.

Analysis

It is not surprising that cities with the greatest number of thefts are often located in states that rank among the top ten for theft. The cities tend to be in states that are near the southern border, possess a major port, are experiencing construction booms, or possess all of these characteristics.

THEFT BY CENSUS POPULATION

Core Base Statistical Area (CBSA)	2013 U.S. Census Population Estimate	2013 Thefts	HE Theft Rate per 10,000 Inhabitants
Orangeburg, SC	90,942	44	4.84
Marshall, MO	23,252	9	3.87
Thomasville, GA	44,869	16	3.57
Moultrie, GA	46,275	16	3.46
Williston, ND	29,595	10	3.38
Hot Springs, AR	97,173	30	3.09
Vicksburg, MS	57,471	17	2.96
Shawnee, OK	71,158	21	2.95
Kinston, NC	58,914	17	2.89

Notes

1. Sources are the total number of thefts reported to NCIC during 2013 and the 2013 U.S. Census report.
2. The term "Core Based Statistical Area" (CBSA) is a collective term for both metro and micro areas. A metro area contains a core urban area population of 50,000 or greater, and a micro area contains a core urban population of at least 10,000 but less than 50,000. Each metro or micro area consists of one or more counties and includes the counties containing the core urban area, as well as any adjacent counties that have a high degree of social and economic integration (as measured by commuting to work) with the urban core.

Analysis

It is not surprising that most of the areas with the highest rates of theft per 10,000 inhabitants are located in the states with the highest numbers of thefts in 2013. What is surprising is that none of the regions in the top ten has a population greater than 100,000. Although the population is small in these regions, more thefts occur per person than in the larger metropolitan areas. The relatively high rate of theft by population in these regions indicates that equipment owners should not be lax with security no matter how remote or loosely populated an area may be. In fact, the data suggests that equipment owners and dealers should be more concerned about equipment theft in regions with smaller populations.

THE COST OF EQUIPMENT THEFT

At present, there is no centralized, accurate, or exhaustive database that includes every loss. NER examines detailed theft reports from a specific area that accurately reports theft—such as a fleet, industry, or region—to make assumptions and develop trends. Then we apply those trends to the entire market share of that specific area to build a national figure. Annual estimates of the cost of equipment theft vary from about $300 million to $1 billion, with most estimates in the range of $400 million.

Notes

1. The estimates don't include the theft of tools or building materials or damage to equipment and premises caused during a theft.
2. The estimates don't include losses from business interruption. Those losses include the cost of rentals, project-delay penalties, and wasted workforce and management time.

Analysis

Several factors contribute to the high level of equipment theft:

- The value of heavy equipment*
- Poor equipment and site security
- Opportunities to sell stolen equipment in the used equipment market
- Low risk of detection and arrest
- Lenient penalties for thieves if prosecuted and convicted

  *The average estimated value of a stolen piece of equipment is $17,400.

RECOVERY STATISTICS
RECOVERY RATES

Low recovery rates make it difficult to draw concrete conclusions from recovery statistics alone. By including information from investigations, such as those in the "Case Studies" section, we can gain an idea of how equipment is stolen, where it goes, and who steals it. The NICB compiled 11,486 reports of stolen machines in 2013. Conversely, in 2013, the NICB reported 2,465 recoveries of equipment listed in the NCIC active theft file. The file includes all active thefts recovered in 2013.

Notes

1. Of the 11,486 reported equipment thefts in 2013, NCIC reported 2,465 recoveries.

2. The recovery rate does not reflect pieces of equipment that law enforcement recovered but did not mark as recovered.
3. The recovery rate does not reflect unreported thefts.

Only 21 percent of stolen equipment was recovered in 2013.

Analysis

Several factors contribute to the low recovery rate of stolen equipment. They are as follows:

- Delays in discovery and reporting of theft
- Inaccurate or nonexistent owner records
- Lack of pre-purchase screening of used equipment
- Limited law enforcement resources dedicated to equipment investigations
- Complexities in equipment numbering systems
- Limited, possibly inaccurate, equipment information in law enforcement systems
- NCIC equipment information reporting errors, in which equipment is erroneously added to the "article file" rather than the "vehicle file"

Comment

The area that needs the most improvement is also the area that promises immediate results: making accurate information available to law enforcement 24 hours a day through NER and the NICB. At a minimum, equipment owners should keep accurate lists of equipment with PIN/serial numbers and submit them to law enforcement, their insurers, and NER as soon they discover a theft. When they purchase equipment, owners should register serial numbers in the NER database, so that the information is available to law enforcement 24 hours a day. In the event of a theft, law enforcement can identify the equipment, even during weekends or at night.

RECOVERY BY STATE

Top Ten States for Equipment Recovery

State	Recoveries
Texas	330
California	302
Florida	179
North Carolina	119
Georgia	100
South Carolina	83
Ohio	81
Tennessee	79
Missouri	78
Oklahoma	72

Notes

1. In 2013, law enforcement recovered most machines in the same state from which they were stolen.
2. The bigger the state and the more demand for equipment within that state, the lower the chance that the equipment will leave the state.
3. If thieves do not sell equipment quickly in the local vicinity, there is a greater chance they will move equipment out of state, especially as more time passes from the date of the theft.
4. Law enforcement is less likely to recover equipment when thieves move it far away, especially out of state. Therefore, more stolen equipment may be moving out of state.

> The top ten states account for 58% of recoveries.

Analysis

1. Lack of screening in the used-equipment market bolsters thieves' confidence. They feel safe selling equipment in neighboring states or even as close as neighboring counties.
2. Recoveries made at ports and borders prove that thieves do export stolen equipment; however, selling stolen equipment within the United States is easy, so the cost of export is worthwhile only when thieves can raise prices abroad or when they steal equipment close to a border.

Comment

In the fight against equipment theft, it is important to act both locally (for example, by circulating theft reports) and nationally (for example, by submitting data to national databases). A key component in the fight is to make it harder for thieves to sell stolen equipment. Buyers of used equipment should check machines at www.IRONcheck. com before buying.

TYPES OF EQUIPMENT RECOVERED

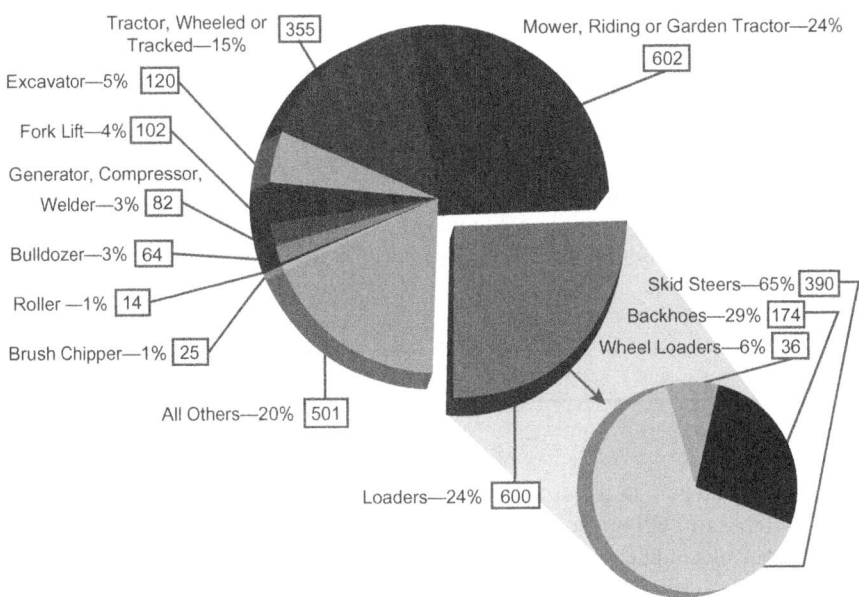

Tractor, Wheeled or Tracked—15% [355]

Excavator—5% [120]

Fork Lift—4% [102]

Generator, Compressor, Welder—3% [82]

Bulldozer—3% [64]

Roller—1% [14]

Brush Chipper—1% [25]

All Others—20% [501]

Loaders—24% [600]

Mower, Riding or Garden Tractor—24% [602]

Skid Steers—65% [390]

Backhoes—29% [174]

Wheel Loaders—6% [36]

Notes

1. The "Loader" category includes all subclasses: front-end, tracked, wheeled, skid steer, and backhoe.
2. The "Excavator" category includes both full-size and compact or mini-excavators.

Analysis

The types of equipment recovered most are usually the types of equipment stolen most. The gap between theft and recovery narrows as NICB training encourages law enforcement to look more closely at the machines stolen more frequently.

RECOVERY BY MANUFACTURER

Manufacturer	Recoveries
John Deere	503
Caterpillar	247
Bobcat	224
Kubota	210
Case	91
Toro	46
Husqvarna	39
New Holland	38
Cub Cadet	34
International	33

Note
1. Source is the total number of recoveries of equipment stolen in 2013.

Analysis
The top five manufacturers account for 52 percent of all recoveries. The make of recovered equipment closely mirrors the make of stolen equipment.

RECOVERY BY MONTH

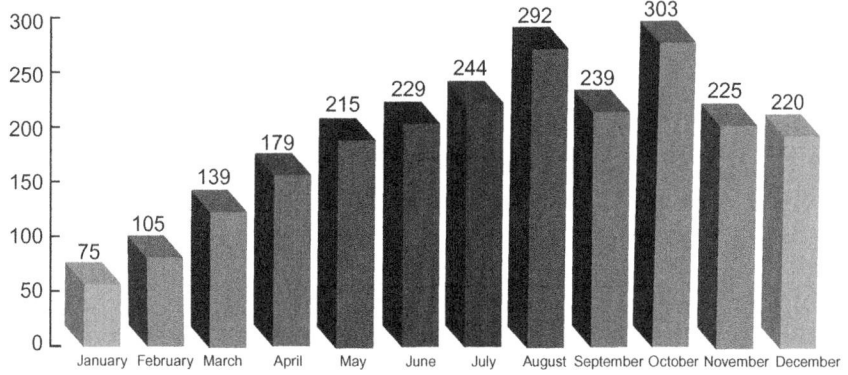

Note
1. Source is the total number of recoveries of equipment stolen in 2013.

Analysis
As the busy construction and farming season slows and jobs near completion, job-sites become safer and more accessible to law enforcement. Larger equipment is generally idle at this point, and even smaller units begin to sit for longer periods as finishing work is done. It is not uncommon for contractors using stolen equipment to abandon it or leave it behind at the end of a job, as maintenance and storage may be more costly than stealing a new machine next year.

MODEL YEAR OF EQUIPMENT RECOVERED

Year	Recoveries
2013	344
2012	316
2011	176
2007	160
2010	159
2005	151
2006	141
2008	135
2004	103
2000	80

Notes

1. Source is the total number of recoveries of equipment stolen in 2013. Each piece of equipment manufactured in 2013 faced potential theft for only part of the year, from the date sold to December 31.
2. Results may be skewed slightly because owners often misstate the date of manufacture. For example, a buyer may list a 2010 model purchased in 2011 as a 2011 model

Analysis

Newer equipment draws more attention from both law enforcement and thieves. It is not uncommon for older equipment to sit unused in lots or yards, but newer equipment is more likely to be noticed as out-of-place by officers.

TOP 10 CITIES FOR EQUIPMENT RECOVERY

City	State	Recoveries
Houston	TX	59
San Diego	CA	34
Miami	FL	28
Bakersfield	CA	25
San Antonio	TX	15
San Bernardino	CA	15
Oklahoma City	OK	12
Anderson	SC	12
Los Angeles	CA	12
Louisville	KY	11
Stockton	CA	11
Fort Lauderdale	FL	11
Riverside	CA	11

Notes

1. Source is the total number of equipment stolen in 2013.
2. If a thief does not sell the equipment immediately in the local area, there is a greater likelihood that, as more time passes, the thief will move equipment out of state and sell it to a purchaser who seems to have no knowledge of the theft.
3. Louisville, KY, Stockton, CA, Fort Lauderdale, FL and Riverside, CA tied for 10th place with eleven recoveries each.

Analysis

Recoveries tend to be localized near high theft areas, suggesting that a good deal of stolen equipment doesn't move far. This may be due to the rules of supply and demand: where there is equipment to steal, there are machines that are needed. Unfortunately, not all high theft areas have high recoveries. Areas with proper funding, training, and dedicated heavy equipment taskforces have much higher recovery rates. It is interesting to note California's significant presence on this list. This state's mandatory statewide registration programs provide law enforcement with many opportunities to access equipment and, therefore, make recoveries.

BY THE NUMBERS
KEY STATISTICS

The following numbers give a snapshot of NER and NICB operations as of December 31, 2013:

21,432,835	Number of ownership records
$9,764,536	Value of items recovered by law enforcement with the help of NICB and NER in 2013
$27,123	Average value of machines recovered by police with NICB and NER assistance
110,197	Theft reports in NER database
11,908	Fleets with equipment registered with NER
4,993	Law enforcement officers trained by NICB on heavy-equipment investigations in 2013
360	Attendees at FBI-LEEDA/NER/NICB Regional Equipment-Theft Summits in 2013
302	Recoveries made by law enforcement with the help of NICB and NER in 2013
25	States in which the NICB conducted training in 2013
48	Number of insurance companies offering incentives to register equipment on NER's database

FINAL NOTES
2013 CASE STUDIES
Emissions Record Leads to Recovery

In May, 2013, NER received a phone call from a detective with the Kern County Sheriff's office asking for help in identifying a stolen Gradall G6-42P Telehandler. The detective was unable to confirm that the equipment was stolen through the National Crime Information Center (NCIC). However, NER had a direct registration record for the Gradall which showed it was originally owned by a rental company that went out of business several years ago. No records were readily available indicating how the assets were liquidated.

Upon further examination, the detective noted that the machine had an owner applied number (OAN) stenciled on its side. The NER analyst assisting the detective believed the supposed OAN was actually a California Air Resources Board (CARB) Registration number. With this new information, NER contacted CARB and inquired about the registration. When it was determined that the CARB record on file did not match original ownership information, the detective was provided with the right contact information of the CARB registrant.

The detective contacted the registrant, who stated the machine was missing. The registrant also provided additional details, which enabled the detective to find the original loss on NCIC.

NICB works with Mexican Authorities to Return Stolen Motor Grader

The National Insurance Crime Bureau (NICB) received a report in June stating that a Caterpillar 140H motor grader worth roughly $140,000 had been located in Saltillo, Coahuila, Mexico. The motor grader was stolen in Dallas, TX in 2011.

Upon further investigation, NICB learned that the individuals in possession of the motor grader were members of the Zetas Drug Cartel. NICB agents informed the Mexican para-military police SWAT team in that area in Mexico and requested their assistance in seizing the machine.

In late September, the Public Ministry in Piedras Negras, Coahuila, Mexico, reported to the Border Auto Theft Information Center (BATIC) and the Texan Department of Safety (DPS) located in El Paso that the Caterpillar motor grader was retrieved and was in government possession.

Once the machine was secure, documentation was provided by the U.S. Consulate Office to finalize the repatriation of the Caterpillar. Even though heavy equipment is not designated in the 1983 U.S.-Mexico Revised Convention on the Repatriation of Stolen Vehicles and Aircraft, NICB's liaison and working relationship with the Mexican Federal Prosecutor's Office facilitated the repatriation of the Caterpillar. In November, the Mexican authorities released the Caterpillar to NICB agents in Piedras Negras, Mexico and moved it to Eagle Pass, TX on the same day.

SUMMARY

Although complete statistics do not exist, it is clear from available data that equipment theft is a serious problem. Estimates derived from data in this year's report suggest the total value of stolen equipment in 2013 is close to $300 million. Those numbers do not include losses from business interruption, such as short-term rental costs, project-delay penalties, and wasted workforce and management time. By frequency of loss, theft is a greater problem than any other type of equipment risk.

Equipment theft levels coincide with the amount of equipment in a particular area. The states with the highest volume of construction and agriculture report the largest number of thefts.

Mobility and value of equipment are the lead contributors to theft. Most thefts are from work sites with little or no security. Given two similar types of machines, a thief will steal the newer one because it is more valuable. In contrast to the automobile industry, there is little difference in equipment security between a new machine and one made several years ago.

Law enforcement recovers as little as 20 percent of stolen equipment. Recovery locations and types closely mirror theft locations and types.

CONCLUSION

Equipment owners and insurers should increase risk management for easily transportable, high-value equipment.

Both equipment security and work site security are necessary to prevent theft. Work site security is especially critical because equipment often sits in areas with little or no physical security.

Officers investigating equipment theft should focus on popular targets and look for red flags, such as unusual location, type of transport, missing decals, altered paint, and especially missing identification plates.

The area that needs the most improvement is also the area that promises immediate results: supplying accurate information to law enforcement 24 hours a day through NER and the NICB.

Glossary

bang for your buck Return on investment (ROI)

bottom line Profit

Bar Association A professional Group for attorneys, barristers and lawyers. Many of the associations provide board certification which allows their members to practice law in their respective jurisdictions.

daily activity report (DAR) The daily document generated by the security officer at the site, detailing the routine and nonemergent events of his or her security activities during working hours

divided site Construction site that is bisected by a roadway, railroad tracks, body of water, or some other geographic barrier

dummy camera A device that appears to be a camera but actually serves no function other than to create an illusion that an area is being surveilled

equipment and materials marshalling The collection and relocation of construction equipment and materials at the end of each workday to protectable grouping compounds

Head's up advanced notice.

Law Society An organization of attorneys that seeks to enhance their professionalism and education.

Looky-Loo A gawker. Someone who looks at something with no constructive reason, and who obstructs or delays the operation.

N/A Not Applicable.

National Equipment Register (NER) An organization that exists separately in the United Kingdom and United States that provides for private registration, tracking, and recovery of heavy construction equipment.

EM Shift – Early Morning The shift between PM (evening) shift and Day shift; generally between 2300 hrs. and 0700 hrs.

proprietary security Security that is provided by the business owner as opposed to a contract security provider

protectable grouping The placing of as many assets as possible into compounds to make them easier to secure. This concept involves secure fencing, good lighting, and an effective camera system combined with an alert, well-trained cadre of security officers.

return on investment (ROI) A reasonable, measurable value for the money one has invested.

Request for Proposal (RFP) In security, the document that the client sends to the construction bidders asking for a business proposal for a construction project

Request for Quote (RFQ) The document that the client sends to the construction bidders asking for a price quote for a construction project. This generally includes a breakdown of costs and profit. Quotes should include the cost for the security program.

rolling the dice Leaving things to chance. In security, this means starting a construction project with no security program in the hope that nothing will happen.

serious incident report (SIR) A report generated by a security officer, detailing the particulars of a significant serious event that occurred during working hours

TER The National Plant and Equipment Register is an organization that is similar to the NER, but operates independently in the UK and Europe.

Vehicle Identification Number (VIN) Unique serial number provided by the manufacturer to identify a specific motorized vehicle

Index

Note: Page numbers followed by "*f*" refer to figures.

Made in United States
Orlando, FL
09 December 2025

74188631R10155